AF512585

BOTANY : AN INTRODUCTION TO PLANT BIOLOGY

BOTANY : AN INTRODUCTION TO PLANT BIOLOGY

Ramesh Bhandari

CYBER TECH PUBLICATIONS
4264/3, Ansari Road, Daryaganj, New Delhi-110002 (India)
Ph.: 011-23244078. Fax: 011-23280028
E-Mail: cyberpublicationsdelhi@yahoo.com

BOTANY : AN INTRODUCTION TO PLANT BIOLOGY

Ramesh Bhandari

© Reserved

First Edition 2009

All rights reserved. No part of this publication may be reproduced, stored in a retrieval system or transmitted, in any form or by any means, electronic, mechanical, photo copying, recording or otherwise, without the prior permission of the publisher.

Published by :

G.S. Rawat for **Cyber Tech Publications**
4264/3, Ansari Road, Daryaganj, New Delhi-110002 (India)
Ph.: 011-23244078. Fax: 011-23280028
E-Mail: cyberpublicationsdelhi@yahoo.com

Printed at : Asian Offset Printers

Preface

Over the past 25 years a major revolution has occurred in biology. Research advances, especially at the molecular level, have permitted exponential increases in our understanding of fundamental life processes. The leading edge of these advances has been in the biomedical disciplines. Human health-related areas have been the major beneficiaries.

There is an increasing realization that other potential beneficiaries of the biological revolution are the agricultural and environmental disciplines. Research opportunities abound in the plant sciences that could make a major impact. Yet, recent reports indicate that plant science has not kept pace with the forefront of biological research. It is time to address this disparity. Modern civilization rests on the successful and sustained cultivation of plants and on the wise use of the biologic and physical resource base on which their cultivation depends. Our knowledge about the world around us is incomplete if we do not include plants in our discoveries, and it is distorted if we do not place sufficient emphasis on plant life. From fundamental discoveries about plant life arise technologies and capabilities that have a wide range of practical applications.

— *Author*

CONTENTS

1

INTRODUCTION

Botany, plant science(s), phytology, or plant biology is a branch of biology and is the scientific study of plant life. Botany covers a wide range of scientific disciplines that study plants, algae, and fungi including: structure, growth, reproduction, metabolism, development, diseases, and chemical properties and evolutionary relationships between the different groups. The word is Greek (*botane*) originally meaning pasture, grass or fodder, perhaps gaining its modern meaning through the idea of a livestock keeper needing to know which plants are safe for livestock to eat.[citation needed] The study of plants and botany began with tribal lore, used to identify edible, medicinal and poisonous plants, making botany one of the oldest sciences. From this ancient interest in plants, the scope of botany has increased to include the study of over 550,000 kinds or species of living organisms.

IMPORTANCE OF BOTANY

HibiscusAs with other life forms in biology, plant life can be studied from different perspectives, from the molecular, genetic and biochemical level through organelles, cells, tissues, organs, individuals, plant populations, and communities of plants. At each of these levels a botanist might be concerned with the classification (taxonomy), structure (anatomy and morphology), or function (physiology) of plant life.

Historically, botany covered all organisms that were not considered to be animals. Some of these organisms are no longer considered to be part of the plant kingdom – these include fungi (studied in mycology), lichens (lichenology), bacteria (bacteriology), viruses (virology) and single-celled algae, which are now grouped as part of the Protista. However, attention is still given to these groups by botanists, and fungi, lichens, bacteria and photosynthetic protists are usually covered in introductory botany courses.

The study of plants has importance for a number of reasons. Plants are a fundamental part of life on Earth. They generate the oxygen, food, fibres, fuel and medicine that allow higher life forms to exist. Plants also absorb carbon dioxide through photosynthesis, a minor greenhouse gas that in large amounts can effect global climate. It is believed that the evolution of plants has changed the global atmosphere of the earth early in the earth's history and paleobotanists study ancient plants in the fossil record. A good understanding of plants is crucial to the future of human societies as it allows us to:

- Produce food to feed an expanding population
- Understand fundamental life processes
- Produce medicine and materials to treat diseases and other ailments
- Understand environmental changes more clearly

HUMAN NUTRITION

Nearly all the food we eat comes (directly and indirectly) from plants like this American long grain rice.Virtually all foods eaten come from plants, either directly from staple foods and other fruit and vegetables, or indirectly through livestock or other animals, which rely on plants for their nutrition.

Plants are the fundamental base of nearly all food chains because they use the energy from the sun and nutrients from the soil and atmosphere and convert them into a form that can be consumed and utilized by animals; this is what ecologists call the first trophic level. Botanists also study how plants produce food we can eat and how to increase yields and therefore their work is important in mankind's ability to feed the world and provide food security for future generations, for example through plant breeding.

Botanists also study weeds, plants which are considered to be a nuisance in a particular location. Weeds are a considerable problem in agriculture, and botany provides some of the basic science used to understand how to minimize 'weed' impact in agriculture and native ecosystems. Ethnobotany is the study of the relationships between plants and people.

Gregor Mendel laid the foundations of modern genetics from his studies of plants.

FUNDAMENTAL LIFE PROCESSES

Plants are convenient organisms in which fundamental life processes (like cell division and protein synthesis for example) can be studied, without the ethical dilemmas of studying animals or humans. The genetic laws of inheritance were discovered in this way by Gregor Mendel, who was studying the way pea shape is inherited. What Mendel learned from studying plants has had far reaching benefits outside of botany. Additionally, Barbara McClintock discovered 'jumping genes' by studying maize. These are a few examples that demonstrate how botanical research has an ongoing relevance to the understanding of fundamental biological processes.

Medicine and Materials

Many medicinal and recreational drugs, like tetrahydrocannabinol, caffeine, and nicotine come directly from the plant kingdom. Others are simple derivatives of botanical natural products; for example aspirin is based on the pain killer salicylic acid which originally came from the bark of willow trees. There may be many novel cures for diseases provided by plants, waiting to be discovered. Popular stimulants like coffee, chocolate, tobacco, and tea also come from plants. Most alcoholic beverages come from fermenting plants such as barley and grapes.

Plants also provide us with many natural materials, such as cotton, wood, paper, linen, hemp, vegetable oils, some types of rope, and rubber. The production of silk would not be possible without the cultivation of the mulberry plant. Sugarcane, rapeseed, soy, cereal grain, biomass crops and other plant products with a high energy content have recently been put to use as sources of biofuels, which are potentially important alternatives to fossil fuels.

Environmental Changes

Plants can also help us understand changes in on our environment in many ways.

- Understanding habitat destruction and species extinction is dependent on an accurate and complete catalog of plant systematics and taxonomy.
- Plant responses to ultraviolet radiation can help us monitor problems like the ozone depletion.
- Analyzing pollen deposited by plants thousands or millions of years ago can help scientists to reconstruct past climates and predict future ones, an essential part of climate change research.
- Recording and analyzing the timing of plant life cycles are important parts of phenology used in climate-change research.
- Lichens, which are sensitive to atmospheric conditions, have been extensively used as pollution indicators.

In many different ways, plants can act a little like the 'miners canary', an early warning system alerting us to important changes in our environment. In addition to these practical and scientific reasons, plants are extremely valuable as recreation for millions of people who enjoy gardening, horticultural and culinary uses of plants every day.

HISTORY

The traditional tools of a botanist.Early examples of plant taxonomy occur in the Rigveda, that divides plants into Vrska (tree), Osadhi (herbs useful to humans) and Virudha (creepers). which are further subdivided. The Atharvaveda divides plants into eight classes, Visakha (spreading branches), Manjari (leaves with long clusters), Sthambini (bushy plants), Prastanavati (which expands); Ekasrnga (those with monopodial growth), Pratanavati (creeping plants), Amsumati (with many stalks), and Kandini (plants with knotty joints).

The Taittiriya Samhita and classifies the plant kingdom into vrksa, vana and druma (trees), visakha (shrubs with spreading branches), sasa (herbs), amsumali (a spreading or deliquescent plant), vratati (climber), stambini (bushy plant), pratanavati (creeper), and alasala (those spreading on the ground).

Manusmriti proposed a classification of plants in eight major categories. Charaka Samhita and Sushruta Samhita and the Vaisesikas also present an elaborate taxonomy.

Parashara, the author of Vrksayurveda (the science of life of trees), classifies plants into Dvimatrka (Dicotyledons) and Ekamatrka (Monocotyledons). These are further classified into Samiganiya (Fabaceae), Puplikagalniya (Rutaceae), Svastikaganiya (Cruciferae), Tripuspaganiya (Cucurbitaceae), Mallikaganiya (Apocynaceae), and Kurcapuspaganiya (Asteraceae).

Among the earliest of botanical work in Europe, written around 300 B.C., are two large treatises by Theophrastus: On the History of Plants (Historia Plantarum) and On the Causes of Plants.

Together these books constitute the most important contribution to botanical science during antiquity and on into the Middle Ages. The Roman medical writer Dioscorides provides important evidence on Greek and Roman knowledge of medicinal plants.

In ancient China, the recorded listing of different plants and herb concoctions for pharmaceutical purposes spans back to at least the Warring States (481 BC-221 BC). Many Chinese writers over the centuries contributed to the written knowledge of herbal pharmaceutics.

There was the Han Dynasty (202 BC-220 AD) written work of the Huangdi Neijing and the famous pharmacologist Zhang Zhongjing of the 2nd century. There was also the 11th century scientists and statesmen Su Song and Shen Kuo, who compiled treatises on herbal medicine and included the use of mineralogy.

Important medieval works of plant physiology include the Prthviniraparyam of Udayana, Nyayavindutika of Dharmottara, Saddarsana-samuccaya of Gunaratna, and Upaskara of Sankaramisra.

In 1665, using an early microscope, Robert Hooke discovered cells in cork, and a short time later in living plant tissue. The German Leonhart Fuchs, the Swiss Conrad von Gesner, and the British authors Nicholas Culpeper and John Gerard published herbals that gave information on the medicinal uses of plants.

Modern Botany

A considerable amount of new knowledge today is being generated from studying model plants like Arabidopsis thaliana. This weedy species in the mustard family was one of the first plants to have its genome sequenced. The sequencing of the rice (Oryza sativa) genome and a large international research community have made rice the de facto cereal/grass/monocot model.

Another grass species, Brachypodium distachyon is also emerging as an experimental model for understanding the genetic, cellular and molecular biology of temperate grasses. Other commercially-important staple foods like wheat, maize, barley, rye, pearl millet and soybean are also having their genomes sequenced.

Some of these are challenging to sequence because they have more than two haploid (n) sets of chromosomes, a condition known as polyploidy, common in the plant kingdom. Chlamydomonas reinhardtii (a single-celled, green alga) is another plant model organism that has been extensively studied and provided important insights into cell biology.

In 1998 the Angiosperm Phylogeny Group published a phylogeny of flowering plants based on an analysis of DNA sequences from most families of flowering plants. As a result of this work, major questions such as which families represent the earliest branches in the genealogy of angiosperms are now understood. Investigating how plant species are related to each other allows botanists to better understand the process of evolution in plants.

Subdisciplines of Botany

Agronomy—Application of plant science to crop production

Bryology—Mosses, liverworts, and hornworts

Economic botany—The place of plants in economics

Ethnobotany—Relationship between humans and plants

Forestry—Forest management and related studies

Horticulture—Cultivated plants

Paleobotany—Fossil plants

Palynology—Pollen and spores

Phycology - Algae

Phytochemistry—Plant secondary chemistry and chemical processes

Phytopathology—Plant diseases

Plant anatomy—Cell and tissue structure

Plant ecology—Role of plants in the environment

Plant genetics—Genetic inheritance in plants

Plant morphology—Structure and life cycles

Plant physiology—Life functions of plants

Plant systematics—Classification and naming of plants

PROTEIN SYNTHESIS IN THE CYTOPLASM

Since gene expression occurs almost exclusively *via* the synthesis of proteins, an understanding of the mechanism of this process is important. Although living organisms make several thousand different proteins with a variety of functions, present evidence suggests that they are all made, by the same basic mechanism, on polysomes. These structures consist of a template RNA molecule, messenger-RNA (mRNA), with several associated ribosomes, and protein synthesis is a complex, energy requiring, multi-enzymic process. A template mechanism of synthesis is dictated by the fact that the protein product has a specificity which depends upon the precise sequence of the constituent amino acid residues in the polypeptide chain. This automatically rules out the normal enzymic method by which most other cell constituents are made, since each enzyme of a set specific for the synthesis of one protein would, in turn, require a further set for its synthesis, *ad infinitum*. However, since some small polypeptides are synthesized by a non-template mechanism, it is possible that some proteins with highly repetitive sequences of amino acid residues, may be synthesized similarly.

Because all organisms probably synthesize proteins by a similar mechanism, the tendency has been to use an organism which is particularly amenable for basic studies. This organism is the prokaryote, *Escherichia coli,* and higher plant material has not been much used because of associated technical difficulties. However, recognition of the importance of the deposition of proteins in plants as our major source of protein food, is leading to an increased spate of work in the desire to understand, and possibly manipulate, this important process in the major food plants.

Proteins are synthesized on free and membrane-bound polysomes in the cytosol or in various sub-cellular organelles, e.g. mitochondria, chloroplasts. Different aspects of the control of protein synthesis are covered in chapters 13 and 15, but it should be re-iterated here that the level of protein in the cell depends on a dynamic balance between synthesis and degradation.

POLYSOMES

Structure of Ribosomes

Ribosomes are classified as eukaryotic or prokaryotic in type, on the basis of:

(a) sensitivity to various antibiotics;

(b) functional interchangeability of soluble factors and ribosomes from different sources; and

(c) structure and sedimentation characteristics, e.g. 70s (prokaryotic) or 80s (eukaryotic).

Bacterial ribosomes, plant chloroplast and mitochondrial ribosomes are classified as prokaryotic, even though they may not always exhibit all of the above characteristics, e.g. plant mitochondrial ribosomes have sedimentation values of about 80s.

Plant and animal cytoplasmic ribosomes are of the eukaryotic type and have sedimentation values of about 80s. This class of ribosomes are heterogeneous in size however, with most of the differences accounted for by differences in the large subunit, the small subunit having been conserved during evolution (Cammarano *et al.,* 1972a,b & c).

Measurements of the size of plant cytoplasmic ribosomes in the electron microscope vary, but they are approximately 25 × 20 nm. Miller *et al.,* (1966) have described them as being acorn-shaped, and several workers have described a cleft in the small subunit. In the rat liver ribosome model of Nonomura *et al.,* (1971) there is a tunnel between the two subunits, directly under the cleft in the small subunit, which is thought to accommodate the mRNA. Ribosomes require Mg^{2+} for structural integrity and dissociate at low Mg^{2+} concentrations into a large and a small subunit with sedimentation values of about 60s and 40s respectively (Ajtkhozhin *et al.,* 1972). There is good evidence that mRNA attaches to the small subunit before the addition of the large subunit completes the ribosome structure, and the different roles of the two subunits are a feature of all the models of ribosome structure and function (see Noll *et al.,* 1973). The 80s ribosome of plants corresponds to a molecular weight of 3.9×10^6 daltons, with molecular weights for the large subunit of 2.4×10^6 daltons and the small subunits, 1.5×10^6 daltons; the small subunit is of the same molecular weight as its mammalian counterpart, the large subunit is smaller. The large subunit of plant ribosomes contains one molecule of 25s RNA, hydrogenbonded to 1 molecule of 5.8s RNA (Payne & Dyer, 1972), and 1 molecule of 5s RNA; the small subunit contains 1 molecule of 18s RNA. The molecular weight of the 25s RNA is 1.3×10^6 daltons, and that of the 18s is 0.7×10^6 daltons (Loening, 1968). The large subunit contains 46% protein, and the small 54%.

The large subunit of pea seedling ribosomes contains 44–45 proteins and the small subunit 32–40 proteins, of which most are basic and of molecular weights between 20×10^3 and 30×10^3 daltons. Some proteins may be represented by more than one copy per ribosome. Proteins extracted from cytoplasmic and chloroplast ribosomes of the same species, show little similarity in two-dimensional electrophoresis or by immunological comparison and are significantly less similar than are the cytoplasmic ribosomes of different species, e.g. beans and wheat (Gualerzi *et al.*, 1974).

Before a complete understanding of the mechanism of protein synthesis is elucidated, it will be necessary to know the spatial relationships and three-dimensional structures of the proteins and the RNAs of the ribosome, as well as those of the associated molecules which also play a part in protein synthesis. Information on the structure of the proteins and the RNA molecules of the *E. coli* ribosome, together with reconstitution experiments, is now well advanced.

Free and Membrane-Bound Polysomes

Ribosomes associate with mRNA to form polysomes, as can be seen in the electron micrograph; the size of the polysome varies according to the length of the mRNA and the number of attached ribosomes. Polysomes are found either free in the cytoplasm or attached to the surface of membranes of the endoplasmic reticulum (ER) and the nucleus. In animal embryonic cells, most of the polysomes are free and engaged in protein synthesis for internal use, whereas in those differentiated animal cells from which large amounts of protein are exported, most of the polysomes are found attached to the ER. Thus, the generalization arose that membrane-bound ribosomes synthesize protein for export and free ribosomes for intracellular use.

It soon became clear however, that membrane-bound ribosomes occur in some tissues which do not export protein, and that in cells where all the protein synthesized is for internal use, different classes of protein are synthesized on the two types of ribosomes. Less information is available for plants. In the developing broadbean seed the highly vacuolate, relatively membrane-free cells of the cotyledon are transformed just prior to the onset of storage protein synthesis, into cells whose cytoplasm is vesicular and in which ER with attached ribosomes is very prominent. This new protein synthesis machinery is assembled at a precise time in the course of seed development for the production, in large amounts, of the storage proteins of the seed (Bailey *et al.*, 1970). Later, during dehydration of the seed, ribosomes become detached and the population of free ribosomes thereby increased. A similar series of events has been recorded for the developing seeds in many other plants.

The way in which ribosomes attach to membranes is not clear; the ribosomes themselves, the membranes and the protein being synthesized, have all been suggested as being involved in the binding. There is evidence from animals that the large subunit is in close contact with the membranes, and Baglioni *et al.* (1971) have postulated that the large subunit binds directly to the membrane, probably at a specific binding site (Sunshine *et al.*, 1971), and that a protein on the membranes is responsible for the attachment (James *et al.*, 1969). Alternatively, it has been suggested that one of the proteins of the large subunit is responsible, whereas other observations suggest that the binding may be dependent on the nascent polypeptide chain. It now seems likelly that, *in vivo*, membrane-bound ribosomes synthesize a different class of protein to free ribosomes, whereas *in vitro*, proteins of both classes are synthesized by both types of ribosomes, suggesting the control is not a function of the ribosome itself.

Isolation and Purification

Ribosomes and polysomes are normally isolated from plants and purified by sedimentation through sucrose cushions as originally described by Wettstein *et al.* (1963).

In a typical procedure, the material is homogenized in 0.25 M sucrose containing 200 mM -Tris-HCl, pH 8.5 at 2°C, 500 mM -KCl and 15 mM -$MgC1_2$, with a Willems Polytron for 3 seconds at a speed setting of 8.

The homo-genate is filtered through Miracloth and Triton X-100 is added to a final concentration of 2% (v/v). The filtrate is centrifuged at $10^4 \times g$ for 10 min at 2°C and the ribosomes and polysomes are recovered from the supernatant by layering over a 3 ml cushion of 0.1 M sucrose in 50 mM -Tris-HCI, pH 8.5, 50 mM -KCl and 10 mM -$MgCl_2$, followed by centrifugation for at least 44 x 10^7 g-min, as described by Leaver & Dyer (1974). The inclusion of the detergent Triton X-100 is necessary since plant materials contain a proportion of polysomes which are membrane-bound and the addition of Triton X-100 solubilizes the endoplasmic reticulum, so releasing them.

Figure: Sucrose density profiles of ribosomes and polysomes from Pisum sativum. Ribosomes recovered after 6 h. centrifugation through 1 M -sucrose cushion. E_{254} = absorbance at 254 nm (From Leaver & Dyer Biochem. J. 144, 165–7, 1974.)

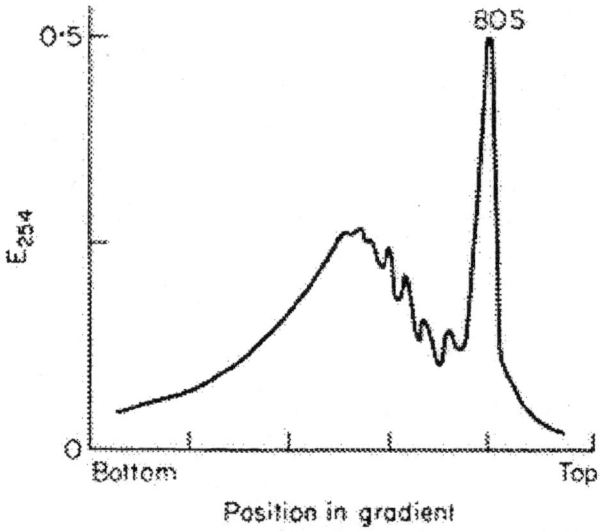

The method gives a preparation of both polysomes and ribosomes and the latter can be removed by centrifugation techniques or, alternatively, the whole preparation can be converted to ribosomes by exposure of the plants to nitrogen gas for at least 1 hour prior to extraction. Since different plants and even different tissues from the same plant contain different amounts of membrane-bound to free polysomes and different free polysomes to free monoribosome ratios, isolation conditions may need to be varied for optimum results with different experimental materials. The situation is further complicated by the fact that plant mitochondrial ribosomes, although not functionally of the 80s type, sediment at 80s (Leaver & Harmy, 1973), and can contaminate cytoplasmic preparations to varying extents, depending on the experimental material. Damage to ribosomes, both structural and by the removal of associated proteins, can occur during isolation and preparation. The extent of this damage can be assessed, to some extent, by extracting the RNA and fractionating it by polyacrylamide gel electrophoresis (see Leaver & Key, 1970).

The Biochemical Mechanism of Protein Synthesis

Protein synthesis, i.e. the assembly of polypeptide chains from their amino acid constituents, occurs on the ribosomes. The information specifying the amino acid sequence of the polypeptide chain is contained in mRNA molecules, which in the case of eukaryotes are complexed with specific proteins *in vivo*. Messenger-like RNA-protein particles called 'informosomes' have been isolated from the cytoplasm and from polysomes of germinating wheat embryos (Ajtkhozhin *et al.*, 1973; Ajtkhozhin & Akhanov, 1974). So far, however, the RNA of these particles has not been definitely established as mRNA. When an mRNA molecule associates with ribosomes, the information it contains is translated; the number of ribosomes involved depends largely upon the size of the polypeptide chain being synthesized, since generally, ribosomes are evenly spaced along mRNA molecules.

The translation process can be arbitrarily split into initiation, elongation, and termination and release. However, prior to the initiation of the process on the ribosome, the twenty different protein amino acids must be activated and attached to specific tRNA molecules. Also, after release of the polypeptide chain from the ribosome, a series of post-translational changes may occur.

Amino Acid Activation and Aminoacyl-tRNA Synthesis

This energy-requiring process is accomplished in the cytosol by enzymes called aminoacyl-tRNA synthetases or ligases. It occurs as a two-stage reaction, both stages being catalysed by the same aminoacyl-tRNA synthetase enzyme. The product of the reaction, aminoacyl-tRNA molecules, contain in the aminoacyl-tRNA bond, sufficient energy for the subsequent formation of peptide bonds to occur spontaneously and, in addition, the anticodon triplet of bases of the tRNA ensures that the amino acid will be located at a particular residue position in the amino acid sequence of the protein product. This reaction, therefore, controls the proportion of free to aminoacylated-tRNAs in the cytoplasm, and theoretically therefore, could affect the rate and/or type of protein being synthesized.

tRNA

The code words (codons) of the genetic code were first established for *E. coil.* This work showed that there were 61 codons shared between twenty protein amino acids, i.e. there is more than one codon for some amino acids and these can attach to more than one tRNA species. Chemically different tRNA species which can be acylated by the same amino acid are called isoacceptor tRNAs.

Figure: The basis of the 'wobble' hypothesis. (X = Any nucleotide; I = Inosine; G = Guanine; U = Uridine; A = Adenine; C = Cytosine;) (From Boulter* et al. *(*Biol. Rev. *47, 113–75, 1972.))

5'-Anticodon base	mRNA codon
I	XXU, XXC, XXA
G	XXU, XXC
U	XXA, XXG
A	XXU
C	XXG

However, it was soon realized that not every codon has a corresponding tRNA with a specific anticodon and that some tRNAs recognize more than one codon. Crick's 'wobble hypothesis' (1966) to account for this, proposed alternative base-pairing to occur between the base in the third position of the mRNA codon and the corresponding base in the tRNA anticodon. Thus, inosine in the 5' position in the anticodon of an aminoacyl-tRNA molecule allows this base to pair with either U, C or A in the third 3' position of a messenger codon. Similarly, with G or U in the 5' position in the anticodon, two codons may be recognized by a single aminoacyl-tRNA, whereas with C or A only a single codon is recognized.

Since alternative base-pairing is only possible with the base in the third position of the codon, codons for the same amino acid which differ in either of the first two base positions, base-pair with different tRNAs. The only example of 'wobble' in the first base of a codon occurs with $tRNA_i^{Met}$, which recognizes both the initiator AUG or GUG codons (see later). Even so, separation of tRNAs by counter current-distribution, MAK columns, BD-cellulose columns and reversed phase chromatography has shown that there are more isoacceptor tRNAs than can be accounted for by the degeneracy of the genetic code. Similarly, isoacceptor tRNAs for amino acids exist in plants; for example, soyabean seedlings have at least six different $tRNA^{Leu}$ species (Anderson & Cherry, 1969). Although few experiments have been carried out, it has been found in every case investigated that plants use the same codon in amino acid assignments as those of *E. coli*. However, plants may not use all of the possible codons for a particular amino acid, since some isoacceptor tRNA species may be absent, or because of the 'wobble' effect; e.g. Caskey *et al.*, (1968) have shown that the isoleucyl tRNA from guinea pig liver recognizes AUU, AUC and AUA codons, whereas the corresponding isoleucyl-tRNA from *E. coli* only recognizes two of these, AUU and AUC. This response to different codons in the two organisms is explained if the liver tRNA had 5' inosine in the anticodon, while *E. coli* tRNA had guanine.

The whole question of the number and the activity of tRNAs in different tissues of plants is complicated by the fact that, (a) it is not certain whether the methods for separating tRNAs are completely effective; (b) it is known that during isolation and purification, partial modification of tRNA molecules can occur, which may affect their charging ability with amino acids or their ability to transfer amino acids to proteins, or both (see Sueoka & Kano-Sueoka, 1970); (c) the relative importance of the three protein synthesizing systems, cytoplasmic, chloroplast and mitochondrial, each with its associated tRNAs and synthetases may differ in different tissues.

Nevertheless, different complements of isoacceptor tRNAs may occur in different tissues and changes in pattern have been demonstrated during development (Bick *et al.*, 1970; Littauer & Inouye, 1973), and it has been suggested that isoacceptor tRNAs are involved in the control of protein synthesis. Garel *et al.*, (1973) have shown that the complement of tRNAs synthesized in the silk gland is related to the mRNAs being translated and they have suggested defining this phenomenon as 'modulated tRNA biosynthesis'.

Aminoacyl-tRNA Synthetases

Aminoacyl-tRNA synthetases have been isolated and purified from a variety of plants (see Boulter 1970; Boulter *et al.*, 1972; Zalik & Jones, 1973), and although a considerable amount of work has been done on their properties, several problems remain. For example, the question of the detailed substrate specificity of different aminoacyl-tRNA synthetases is still under investigation. Since activation is a two-step process, both amino acid and tRNA specificity have to be considered.

The synthetase is normally absolutely specific towards the tRNA; few mismatches have been reported in homologous systems (Arca *et al.*, 1967, 1968). However, a synthetase specific for one amino acid can, in some instances, activate to a lesser extent another amino acid. Even so, synthetases are remarkably amino acid specific.

A second amino acid attached to a tRNA molecule normally specific for another amino acid, will be located in the polypeptide chain according to the tRNA anticodon, i.e. where the first amino acid would have been placed. With regard to amino acids not normally found in the cell, specificity is not always so pronounced, e.g. azetidine-2-carboxylic acid is activated and transferred to $tRNA^{Pro}$ by the proline enzyme of mung bean (Peterson & Fowden, 1965), whereas *Polygonatum multiflorum,* which contains a high level of this amino acid in nature, contains a prolyl-tRNA synthetase which will not activate it. In *Mimosa* and *Leucaena,* where mimosine, an analogue of phenylalanine, occurs naturally and where it is non-toxic, the phenylalanyl-tRNA synthetase discriminates against mimosine. In several other species where mimosine is toxic, it is activated by the phenylalanine-tRNA synthetase but the enzyme does not attach it to $tRNA^{Phe}$ (Smith & Fowden, 1968).

It is not clear if there is one or more than one synthetase whenever there are a number of isoacceptor tRNAs. In soybean seedlings where there are at least three different leucyl aminoacyl-tRNA synthetases (Kanabus & Cherry, 1971) which have different specificities towards the six leucine isoacceptor tRNAs, it would appear that there is more than one cytoplasmic leucyl-tRNA synthetase.

However, in some instances, a single aminoacyl-tRNA synthetase can recognize different isoacceptor tRNAs. When both cytoplasmic and chloroplast tRNA species are present, there are probably at least two sets of aminoacyl-tRNA synthetases.

In light-grown *Euglena* for example (Reger *et al.*, 1970; Krauspe & Parthier, 1973), there are two aminoacyl-tRNA synthetases for each of phenylalanine and leucine, whereas dark-grown *Euglena,* which is without chloroplasts, contains only one; there are various reports of chloroplast aminoacyl-tRNA synthetases which only acylate chloroplast tRNAs and not their cytoplasmic counterparts and *vice versa.* However, in heterologous systems such as chloroplast tRNA plus cytoplasmic activating enzymes, specificity may range from none to complete (Burkard *et al.,* 1970; Boulter *et al.,* 1972), although the biological significance of these findings is not clear.

There is evidence that, as well as the tRNAs, the complement of aminoacyltRNA synthetases may change during the life cycle of plants (Bick & Strehler, 1971).

Translation of mRNA

Initiation of the Polypeptide Chain

Initiation of protein synthesis on plant cytoplasmic ribosomes starts with the assembly of a mRNA•40s subunit complex which requires ATP for its formation and this step is followed by the addition of methionyl-tRNA$_i$ to form a mRNA•40s subunit•methionyl-tRNA$_i$ complex. The process requires GTP and at least two initiation factors (Weeks *et al.*, 1972; Marcus *et al.*, 1973). Initiation is completed by the addition of the 60s subunit and the 80s ribosome so formed is then able to accept the next aminoacyl-tRNA, thus starting the elongation process. The tRNA$_i$^{Met} is a special tRNA charged with unformylated methionine, which can enter the so-called 'P' site of the ribosome complex and base-pair with the initiator codon, AUG or GUG in the mRNA.

Although mRNAs are translated from the 5'® 3' end, evidence from sequence studies of bacteriophage messages shows that the initiator triplet is not at the 5' terminus, but is set a considerable number of nucleotides in. For example, the first initiator codon of bacteriophage R17 RNA is preceded by 91 nucleotides which are not translated (Adams & Cory, 1970). In the case of polycistronic messengers, one or more initiator triplets will occur intramolecularly.

Evidence that plants have a special type of initiator tRNA, tRNA$_i$ ^{Met}, comes from the wheatgerm system (Leis & Keller, 1970; Marcus *et al.*, 1970a; Tarrago *et al.*, 1970; Ghosh *et al.*, 1971), and from *Vicia faba* (Yarwood *et al.*, 1971a), where it has been shown that two major and one minor tRNA^{Met} species are present.

The minor and one of the major tRNA^{Met} species function as chain initiators as shown by AUG-dependent binding (Tarrago *et al.*, 1970; Yarwood *et al.*, 1971a), and by N-terminal analyses of either the product of tobacco mosaic viral RNA-directed (Marcus *et al.*, 1970a) or poly-AUG, poly-GU and endogenous messenger-directed incorporation (Yarwood *et al.*, 1971a). Neither of the major tRNA^{Met} species from either wheat or beans is formylated, in contrast to the initiator tRNAi^{Met} of prokaryotes, and it is presumed that these are the cytoplasmic tRNA^{Met} s for initiator and internal methionine residues. The minor, formylatable $tRNA^{Met}$, is presumed to be the initiator of chloroplast protein synthesis.

The hydrolysis of ATP precedes the formation of the first peptide bond and may be required for recycling of the methionyl-$tRNA_i$ binding factor. The requirement for ATP has also been demonstrated in the rabbit globin synthesizing system (Schreier & Staehelin, 1973), but not in those of prokaryotes. The order in which the initiator tRNA and mRNA bind to the small subunit in mammalian and prokaryote systems is in dispute; the order suggested above for wheatgerm by Marcus and his coworkers, therefore, whilst agreeing with some workers on those systems disagrees in particular with the suggestion of Legon *et al.*, (1973) and Schreier and Staehelin (1973) that the initiator binds prior to the mRNA in rabbit globin synthesis.

Knowledge of the two plant initiator factors involved in binding methionyl-$tRNA_i$ and mRNA is much less complete than for the similar factors in bacterial and mammalian systems; details of the latter are, therefore, relevant. In bacteria there are three factors, IF-1, IF-2 and IF-3 involved in the initiation process. Factor IF-1 has a molecular weight of 9,400 daltons and is the most basic of the three proteins.

Factor IF-2 has been shown to consist of two sub-components, both active, with molecular weights of $9 \times 10^3 - 100 \times 10^3$ daltons and 80×10^3 daltons although the latter is probably derived from the former by proteolysis during purification. Factor IF-3 is a protein or group of proteins with molecular weight(s) of $21.5 \times 10^3 - 23.5 \times 10^3$ daltons (Grunberg-Manago *et al.*, 1973). The three factors attach to the small subunit; IF-2 binds GTP and formyl-methionyl-RNA, while IF-3 binds mRNA. Factor IF-I increases the affinity of factors IF-2 and IF-3 for the 30s subunit, and is also necessary for the release of IF-2 from the ribosome.

The order of the attachment of the different factors to the small subunit and the order of the binding of mRNA and acylated initiator tRNA, is still a matter of debate. Some results suggest that IF-1 is attached before IF-2 and IF-3, whilst others indicate that it can only be attached after. Similarly, it is possible that an unstable intermediary complex consisting of formyl-methionyl-tRNA$_i$ •30s subunit•initiation factors is formed, i.e. that mRNA is bound after initiator-tRNA, or that a 30s subunit•mRNA•IF-3 complex is formed, which is then stabilized by the addition of formyl-methionyl-tRNAi and IF- I and IF-2. The next step in the process, the addition of the large subunit, does not require GTP, but GTP hydrolysis is required to release IF-2 from the 70s ribosome complex. The role of GTP hydrolysis in initiation is still not fully understood, but it is not only required for the release of IF-2. Factor IF-3 is not only required for the binding of natural mRNAs, but also functions later either to dissociate the subunits after chain termination, or to keep dissociated subunits apart until another cycle of initiation is instituted. This activity of the IF-3 factor is referred to as its DF (ribosome dissociation factor) activity.

Several proteins which form an integral part of the small subunit have also been shown to be involved in initiation. Small subunit proteins, s1, s11, s13, s19, participate in IF-2 binding and s1, s11, s12, s13, s14, in IF-3 binding; s12 is responsible for recognition of mRNA (Haselkorn & Rothman-Denes, 1973; Anderson *et al.*, 1974). Shine and Dalgarno (1974) have proposed a model of how the large subunit.initiation factor complex may recognize and attach mRNA. It would appear that the sequence (5') GGAGGU (3') is present in the same relative position with respect to the first translatable AUG triplet in all prokaryotic messengers so far analyzed. Furthermore, the 3' proximal end of 16s RNA has the sequence GAUC<u>ACCUCCU</u> UA (OH), so that the underlined nucleotides could potentially base-pair with the GGAGGU sequence in the mRNA.

Further complexity of the initiation process is indicated by the work ofSchreier and Staehelin (1973), who have characterized five initiation factors (IF-E_1 to IF-E_5) for mammalian protein synthesis. Their results with rabbit globin mRNA are consistent with the formation initially of a methionyl-tRNA$_1$ •IF-E_2 •GTP complex, which is bound independently of mRNA to the 40s ribosome subunit by IF-E_3. Subsequent to this, mRNA and the 60s subunit are joined by the cooperative action of IF-E_4, ATP, IF-E_1 and IF-

E_5. The binding of natural mRNA requires IF-E_4 and ATP. IF-E_5 promotes the joining of the 40s complex with the 60s subunit and IF-E_1 inhibits complex formation in the absence of mRNA binding. However, it has been suggested that the apparent need for additional factors results from deproteinization of ribosomal subunits and mRNAs, i.e. the requirement may be for structural proteins rather than initiation factors. The little information available suggests that eukaryotic factors are not exchangeable with their prokaryotic counterparts.

Table. Probable relationship between factors.

	Eukaryotes			Prokaryotes
Schreier & Staehelin (1973)		Anderson	Anderson *et al.* (1974)	Haselkorn & Rothman-Denes (1973)
IF-E_2		M_1	EIF-1	IF-2
IF-E_3		M_3	EIF-3	IF-3
IF-E_4		M_{2B}		
IF-E_5		M_{2A}		

Columns 1–3 are the various nomenclature designations given to the equivalent factors of mammalian eukaryotic systems, the last column to prokaryotes.
For original literature see references in Schreier & Staehelin (1973).

Factor EIF-3, contrary to its prokaryotic counterpart, consists of a number of different sized polypeptide chains; the question as to whether there are different EIF-3s or whether selectivity is controlled by additional mRNA specific factors, is still unresolved. A similar uncertainty to that described for the prokaryotic system surrounds the question as to whether or not methionyl-tRNA$_i$ binds to the 40s subunit prior to mRNA.

The initiation process is extremely complex and still not fully understood. A variety of proteins are involved, some being ribosomal structural proteins and others proteins which recycle during the process. These proteins interact, changing the conformation of the ribosome and thereby allowing the different steps of initiation to proceed in a sequential and orderly manner. Different mRNAs probably differ in their rate of attachment to the ribosome and/or in their efficiency in other stages of the initiation process, so affecting the frequency and rate of translation. With proteins which contain prosthetic groups their absence may inhibit initiation, since it has been shown that lack of haemin in globin synthesis, results in the formation of an inhibitor of the binding of methionyl-tRNA$_i$ to 40s subunits (Adamson *et al.*, 1972; Gross & Rabinovitz, 1973; Legon *et al.*, 1973). Furthermore, a variety of mRNA specific factors have been proposed called 'interference' or 'i-factors', which affect the specificity of the initiation factor IF-3 (Groner *et al.*, 1972).

Thus, Strycharz *et al.* (1973) have identified a supernatant factor in Krebs II ascites cells, which is specifically required for the initiation of the translation of encephalomyocorditis viral RNA. However, Lodish (1974), from a kinetic analysis of protein synthesis, has proposed a model which precludes the necessity for mRNA specific factors. He points out that these have often been observed in systems which are less active than the corresponding *in vivo* system, and that the three eukaryotic systems available which translate exogenous mRNAs efficiently *in vitro,* do so with a variety of mRNAs.

Elongation of the Polypeptide Chain

The elongation process starts once the 80s plant ribosome containing mRNA and methionyl-tRNA$_i$, has been assembled and takes place by the following repeated cycle of events, each cycle being separated into:

(a) codon-directed binding of aminoacyl-tRNA;

(b) peptide bond formation; and

(c) translocation.

The ribosome has two binding sites; a 'P' site for the binding of methionyl-tRNA$_i$ and an 'A' site where all other aminoacyl-tRNAs bind. After the initiation complex has been formed, the next aminoacyl-tRNA, i.e. the one carrying the anticodon to the next codon of the mRNA, binds at the 'A' site. Once the aminoacyl-tRNA has been bound in the 'A' site, a peptide bond is formed by a peptidyl-transferase enzyme, which is part of the large subunit, such that the 'A' site now carries the aminoacyl-tRNA joined to methionine by a peptide bond. The 'P' site now carries the deacylated $tRNA_i^{Met}$, which is subsequently removed leaving an open 'P' site, and the peptidyl-tRNA moves from the 'A' to the 'P' site, so moving the message relative to the ribosome. The 'A' site is now empty and is ready for the next aminoacyl-tRNA to enter according to the next codon in the message. This cycle of reactions requires K^+, Mg^{2+}, GTP and various elongation factors, proteins, which are found in the soluble fraction of cell homogenates.

Figure: Polypeptide chain elongation in prokaryotes. Ribosomes cover more than two codons on mRNA. Reaction I ® II is catalysed by peptidyl transferase. (= tRNA, $(AA)_x$ = peptide with N-terminal amino acid x. AA_{x+1} = amino acid x+ 1.) (Modified from Haselkorn & Rothman-Denes (1973).)

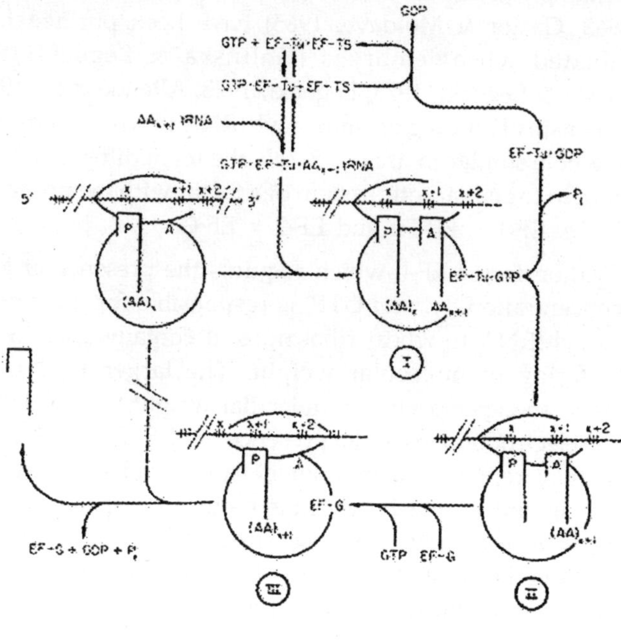

Our knowledge of polypeptide chain elongation is most complete with *E. coli,* where three elongation factors are known, EF-Tu, EF-Ts and EF-G. Factor EF-Tu binds stepwise with GTP and aminoacyl-tRNA and this ternary complex transfers the aminoacyl-tRNA to the 'A' site of the ribosome, releasing an EF-Tu•GDP complex and inorganic phosphate. Factor EF-Ts displaces GDP to form the complex EF-Ts•EF-Tu, from which GTP displaces EF-Ts to regenerate EF-Tu•GTP. The third elongation factor, EF-G, and free GTP, are responsible for the translocation of the peptidyl tRNA from the A' to the 'P' site with the prior removal of the deacylated tRNA from the 'P' site; GTP is hydrolyzed during the process. The elongation factors, like the initiation factors, recycle during protein synthesis. The protein chain grows from the N-terminal end, and probably starts to fold into its three-dimensional conformation whilst the process of elongation is proceeding.

Fewer details of the elongation factors are available in eukaryotic systems. However, two elongation factors, designated as the binding enzyme EF-1 and the translocase EF-2 (Hardesty *et al.,* 1963; Gasior & Moldave, 1965) have been purified from ungerminated wheat embryos (Golinska & Legocki 1973; Twardowski & Legocki, 1973; Legocki, 1973; Allende *et al.,* 1973). The mechanism of binding of aminoacyl-tRNA to wheat ribosomes is on the whole similar to that in *E. coli,* the main difference being the absence in the eukaryotic system of a factor with the properties of EF-Ts, i.e. EF-1 = EF-Tu and EF-2 = EF-G.

Elongation factor EF-I, which requires the presence of Mg^{2+} at low concentration, K^+ and GTP, is responsible for the binding of aminoacyl-tRNA to wheat ribosomes. It contains three active forms differing in molecular weight. The larger EF-1 forms dissociate to a single species of molecular weight 6×10^4 daltons during gel electrophoresis in the presence of SDS. The three forms observed may represent a monomer, trimer and tetramer of a single protein unit, of which the trimer seems to be the most stable form. Some experimental evidence supports the view that the light species is the active form of the EF-1 enzyme. For example, in calf brain and wheat embryo, it has been shown that the interaction of EF-1 with GTP and aminoacyl-tRNA yields a ternary complex which contains the light form of the enzyme, i.e. molecular weight 5 × 10^4 to 6 × 10^4 daltons (Moon *et al.*, 1972; Legocki, 1973). It is suggested that the amino group of the esterified amino acid plays an important role in aminoacyl-tRNA binding (Jerez *et al.*, 1969) and this differs from the bacterial system where EF-Tu can react with the deaminated product of phenylalanyl-tRNA, phenyl-lactyl-tRNA (Fahnestock *et al.*, 1972).

The second elongation factor, EF-2, is involved in translocation of the peptidyl-tRNA from the 'A' site to the 'P' site during elongation. It is a protein of molecular weight 7 × 10^4 daltons and requires thiol compounds for activity (Twardowski & Legocki, 1973).

Ribosomes and elongation factors from the cytoplasm of eukaryotic organisms are not functionally interchangeable with their *E. coli* counterparts (Krisko *et al.*, 1969). Furthermore, Perani *et al.*, (1971) have shown absolute ribosome specificity, either for 70s or 80s ribosomes, of two sets of elongation factors, T and G (EF-1 and EF-2), isolated from yeast, and for factor EF-2 isolated from the alga, *Prototheca zopfii.* It is generally accepted that elongation factors within each ribosomal type (70s or 80s) are exchangeable between eukaryotes (Ciferri, 1972).

Termination and Release of the Polypeptide Chain

No reliable information is available on chain termination in plants, most of the work having been done with *E. coli.* However, as there are some differences between the prokaryote and mammalian eukaryote systems, these will be mentioned briefly.. In *E. coli* the elongation cycle is repeated until certain termination codons on the mRNA come into the 'A' site; these codons are UAG, UAA and UGA. At least three release factors, proteins, which recognize the terminator codons and bring about the release of the completed polypeptide chain from the tRNA, have been identified (Caskey *et al.*, 1969; Capecchi & Klein, 1969; Milman *et al.*, 1969). Originally designated R-1, R-2 and S these are now called RF-1, RF-2 and RF-3. RF-1 recognizes UGG or UAG, RF-2 recognizes UAA or UGA, and RF-3 affects the rate of release of the polypeptide chain. In mammalian systems, only a single release factor which responds to all three codons has been identified, and furthermore the prokaryotic termination requirement for GTP appears to be absent (Haselkorn & Rothman-Denes, 1973).

In addition, several other termination factors have been proposed from different organisms (see Haselkorn & Rothman-Denes, 1973). Although little is known about chain termination in plants, the presumed universality of the code suggests that similar terminator triplets and proteins are involved in plants also.

Originally, it was thought that on chain termination ribosomes dissociated into subunits and that the free ribosomes found in the cell were inactive in further protein synthesis. Noll *et al.* (1973) have presented evidence that at least some of the free ribosomes found in the cell interact with a factor and partially dissociate, so that the small subunit with bound IF-1, IF-2 and IF-3, can initiate protein synthesis.

In conclusion, it may be said that small differences exist between the mechanism of protein synthesis in the pro- and eukaryotic systems, e.g. the initiator tRNA, the involvement of GTP in different steps, as well as in the possible number of elongation factors. In view of the technical difficulties involved, work on the detailed mechanisms of eukaryotic systems could be considered hardly worthwhile. However, it is possible that translation level controls are much more important in eukaryotes than they are in prokaryotes. This possibility, together with the importance of plant proteins as a source of food, are justification enough for an attempt to characterize the process in detail in plants and, thereby develop suitable assay systems.

Cell-Free Systems

In order to gain detailed knowledge of the mechanism of protein synthesis in plants, it is necessary to develop *in vitro* assay systems. Such systems must satisfy the following criteria:

(a) be dependent on the exogenous component(s) to be tested;

(b) be efficient, i.e. of comparable activity to *in vivo* rates, in order to allow quantitative analysis;

(c) the product, whose synthesis is monitored, must be clearly defined.

Two basic types of cell-free systems are used: fractionated systems, in which enzymes and components are isolated, purified and then reconstituted into an assay system, and unfractionated systems, in which the cells are broken open and cell-debris, nuclei and mitochondria removed by centrifugation leaving the ribosomes and the various components for protein synthesis in the supernatant, which is then used as the assay system. If all of the components of the assay are isolated from one organism, the system is said to be homologous as opposed to heterologous when they are not.

In addition to the fractionated *E. coli* prokaryotic system, there are four fractionated eukaryotic systems which satisfy the above criteria. These are the rabbit reticulocyte (Lockard & Lingrel, 1969; Gilbert & Anderson, 1970), the mammalian liver (Prichard *et al.*, 1971), the Krebs II ascites cell (Mathews & Korner, 1970), and the only plant system, that of wheatgerm (Allende & Bravo, 1966; Allende, 1970; Leis & Keller, 1970; Marcus *et al.*, 1970a,b; Tarrago *et al.*, 1970; Ghosh *et al.*, 1971; Legocki & Marcus, 1970; Klein *et al.*, 1972; Lundquist *et al.*, 1972). Other fractionated cell-free systems from plants, such as those from developing legume seeds (Gumilevskaya *et al.*, 1971; Payne *et al.*, 1971a,b; Yarwood *et al.*, 1971a,b; Beevers & Poulson, 1972; Wells & Beevers, 1973), are not well characterized and assays often lack quantitative accuracy. This is because in isolating the constituent enzymes and components, damage is caused in breaking the cell wall by hydrodynamic sheer, by cell vacuoles breaking and releasing acids, phenolics, tannins and other substances, and by activation of proteases and nucleases (Payne & Boulter, 1974). If the preparations used contain several components and structural damage to ribosomes has occurred, interpretation of the results may be qualitatively ambiguous, since enzyme preparations may contain proteins needed for the reconstitution of protein-leached ribosomes.

Unfractionated cell-free systems cannot be made from tissues which have a high nuclease or protease activity, and until better methods for the inhibition of degradative enzymes become available, most unfractionated systems from plants will be of limited value. Recently, Davies *et al.* (1972) have used media of high ionic strength and high pH with some success, and Gray and Kekwick (1973) have developed a system from pea seedlings using 0.2 mM vanadyl sulphate as an inhibitor.

This system has been shown to synthesize the small subunit of ribulose bisphosphate carboxylase (Fraction 1 protein), since the tryptic peptides of the *in vitro* product were similar to those of the naturally occurring protein. The identity of the product was also proved by immunoprecipitation. The requirement of product identification is essential in complete cell-free assays.

The wheatgerm system is by far the most successful unfractionated cell-free system from plants. Its preparation is as follows (Marcus *et al.*, 1974):

Dry wheat embryos are ground thoroughly with a small amount of sand in a precooled mortar in a total volume of 3.3 ml of 90 mM KC1, 2 mM $CaCl_2$, 1 mM Mg $(Ac)_2$, 6 mM $KHCO_3$. The embryos are initially ground in 1.0 ml with 0.5 and 1.8 ml increments added subsequently.

The slurry is then centrifuged for 10 minutes at 23,500 × *g* and the supernatant is removed with a Pasteur pipette, taking care to leave behind as much as possible of the upper lipid layer. Just prior to use, 0.5–3.0 ml are dialyzed against 500 ml of 1 mM Tris–acetate, pH 7.6, 50 mM KCl, 2 mM $Mg(Ac)_2$, 4 mM 2-mercapto-ethanol for 1.75 hours; this preparation is termed S23.

The wheatgerm system is attractive in spite of needing some additions, (e.g. tRNA), since it shows great promise as a 'translation' system for various mRNAs. This is an important development as the isolation of mRNAs from plants is now feasible using binding to oligo-dT columns (since a proportion of eukaryote messengers contain a poly-A sequence), by gradient centrifugation and by immunoprecipitation of polysomes.

Four RNAs of brome mosaic virus (BMW) induce amino acid incorporation into proteins when used as messengers in the wheatgerm system. RNA4 is translated with an efficiency comparable to that of bacteriophage RNA in *E. coli* extracts. The product is homogeneous and indistinguishable from the coat protein of BMV (Shih & Kaesberg, 1973). Satellite tobacco necrosis virus RNA has also been translated in this system and the product shown to be coat protein (Klein *et al.*, 1972). Several natural eukaryotic messengers have also been translated in the wheatgerm system, e.g. rabbit globin mRNA (Efron & Marcus, 1973) and leghaemoglobin RNA (Verma *et al.*, 1974). The latter is the only naturally occurring plant messenger to be isolated so far. It is a poly A-containing 9s–12s mRNA which was isolated from soybean root nodule polysomes. When used to programme the wheatgerm system by Verma *et al.*, (1974), the product was identified serologically as mainly leghaemoglobin S (Lb_S) with a little leghaemoglobin F (Lb_F); there are two types of leghaemoglobin found in soybean nodules and the reason for the preferential synthesis of the Lb_S *in vitro*, is not understood. Factors other than those already present in the wheatgerm system were not required, and there was no need for the addition of heme.

There are several unfractionated cell-free systems from animals (Haselkorn & Rothman-Denes, 1973). Of particular interest is that of the *Xenopus* oocyte since intact eggs are used to translate exogenous messages (Gurdon *et al.*, 1971). Natural plant messages have not been used but there is evidence that tobacco mosaic virus RNA can be translated (Knowland, 1973). Coat protein is not synthesized, however, and the main product is a polypeptide with molecular weight 14 × 104 daltons. Although no function has been assigned to this protein, it is known that plant cells infected with TMV also make a protein of the same molecular weight (J. Gurdon, personal communication).

The use of heterologous systems assumes the interchangeability of the various enzymes and components between different systems and as mentioned previously, it is generally accepted that this is not usually possible between prokaryotic and eukaryotic organisms. Generally, attempts to demonstrate interchangeability between different eukaryotic systems have been positive, but several workers have suggested that this is because discrimination can only be seen under optimum conditions. The general conclusion is that tissue and organism specificity may occur but only in exceptional cases (Mathews, 1973).

Inhibitors

Several antibiotics and other inhibitors block various steps in protein synthesis, and are, therefore, extremely useful in establishing the mechanism of the process (Pestka, 1974). Table 10.2 lists the more common inhibitors and their sites of action; many of these have been used with plants.

It is important to realize that the concentration of the inhibitor is often critical for the production of a specific effect; it is essential to establish the conditions for the correct use of an inhibitor with each tissue and to maintain effective controls. It has been suggested by Glazer and Sartorelli (1972) that in rat liver, membrane-bound 80s ribosomes are more susceptible to a range of inhibitors than are the 80s ribosomes free in the cytosol.

Post-Translational Changes

Changes which take place after the synthesis of the polypeptide chain are called post-translational. The most common of these are concerned with the many proteins which consist of more than one polypeptide chain.

Since there is no evidence that unpartnered subunits occur in any quantity, machinery must exist in the cell to ensure that approximately correct numbers of each constituent polypeptide chain are synthesized and correctly assembled, although little is known about the mechanisms. Some mRNAs may be polycistronic but in other cases mRNAs for each polypeptide chain are translated independently. Self-assembly by interaction at specific sites on different polypeptide chains may occur.

Several examples are known where some subunits of a protein are made in the cytoplasm and others on the chloroplast or mitochondria. For example, ribulose bisphosphate carboxylase is a large molecule (also known as Fraction I protein) consisting of two types of subunit; the small subunits are synthesized on 80s cytoplasmic ribosomes, whilst the large subunits are synthesized on 70s chloroplast ribosomes and assembly of the two types to form the complete molecule occurs in the chloroplast.

Post-translational changes also occur by the addition of prosthetic groups or by modification of the polypeptide chain. Such modifications may be the formation of disulphide linkages, the removal of N-terminal portions, the removal of intramolecular segments, the derivation of amino acids and the addition of carbohydrate or lipid units to form glycoproteins and lipoproteins respectively.

These changes are all brought about by specific enzymes and probably after the polypeptide chain has left the ribosome.

The formation of disulphide linkages is often part of the assembly of multichain proteins, e.g. legumin in *Vicia faba* (Wright & Boulter, 1974), although the enzymes involved have not been demonstrated from plants.

Few plant proteins have methionine at their N-terminus when isolated, although this amino acid must be presumed to be the N-terminus when synthesized, and, therefore, removal of one or more residues at the N-terminal region must have occurred subsequently.

Enzymes for this purpose have been demonstrated in bacteria but not in plants, and it is possible that the process in plants is non-specific and mediated by leucine aminopeptidase, a ubiquitous plant enzyme which removes amino acids from the N-terminus of proteins and may continue to do so until the conformation of the protein inhibits its activity.

The removal of intramolecular segments of plant proteins similar to the situation with proinsulin in animals, has not been demonstrated, but there is no reason to suppose that it does not occur.

Modification of individual amino acids is quite common in plant proteins; examples are, *S* -N-trimethyllysine in some cytochromes *c* (Ramshaw *et al.*, 1974), and hydroxyproline in extensin, a protein of cell walls (Lamport, 1965). Where investigated, the protein when synthesized contained the parent amino acid, the substituted amino acid arising secondarily, probably after the protein had left the ribosome.

Some residue positions of a particular amino acid may be substituted and others not, though the specificity of the changes is not understood.

Synthesis of the protein moiety of glycoproteins takes place on the membrane-bound ribosomes of the ER, but the sugar groups are added sequentially at different sites in the cell.

Those closely linked to the polypeptide chain such as glucosamine and mannose, are added immediately after release from the ribosome or in the ER itself, other more terminal groups may be added later, some in the Golgi apparatus (Whaley *et al.*, 1972).

PROTEIN SYNTHESIS IN VIVO

The rate of protein synthesis *in vivo* can be measured by determining the amount of protein in a plant, organ or tissue at two different times; this will give the net rate since proteins 'turnover' and the amount of protein found would represent the balance between synthesis and degradation (Huffaker & Petersen, 1974).

Various equations describing the relationship between the synthesis and degradation of protein have been proposed but there are many technical and interpretive difficulties in this type of experiment. The turnover rate for an individual protein is not the same as the value obtained for the total protein and although the latter may be of interest agriculturally, most biochemical and physiological studies require information on individual proteins.

This information has only been obtained so far with a few plant proteins, although most are thought to turnover. Possible exceptions are storage proteins which are sequestered from the metabolically active cytoplasm and stored in membrane-bound protein bodies until subsequently required at a more active phase of the life-cycle.

A variety of studies using acrylamide gel electrophoresis has shown that the protein complement of organs and tissues changes during the life-cycle of a plant, and also in response to environmental conditions. Organ-specific proteins have been demonstrated serologically and often one or a few proteins may predominate in an organ, e.g. ribulose bisphosphate carboxylase, which often accounts for 50% of the protein of leaves.

The changing complement of proteins during development is due to changes in the type or rate of protein synthesis and/or degradation. For example, Beevers and Poulson (1972) followed changes in the protein content of pea cotyledons during the period 9–33 days after flowering. Initially, protein content increased gradually, then rapidly between 21–27 days to decline once more as the seed dehydrated and matured.

Incorporation experiments with ^{14}C-leucine indicated that albumin proteins were synthesized early in cotyledon development whilst globulin synthesis predominated with increasing maturity.

During the period of rapid protein synthesis, greater amounts and a higher percentage of polysomes to monosomes were extracted than at other stages in the development of the seed. However, unlike in bacteria, where induced enzyme synthesis is widespread and the product is not 'turned over' but diluted out by a rapid rate of cell division, plants have a much slower rate of cell division and protein degradation appears to be a common method for regulating the amount, and hence the activity of proteins.

The relative amounts of different seed proteins synthesized may vary considerably in different varieties and breeding lines of a crop; for example, in maize where the major storage proteins of the seeds are prolamins and glutelins normal maize contains a relatively high proportion of prolamin, whereas in the mutant Opaque 2, the proportion of prolamin is greatly reduced. This is of particular significance since prolamin is very deficient in the essential amino acid lysine, so that seed meals obtained from Opaque 2 have a higher content of lysine and are consequently more nutritious than those from normal maize.

Protein synthesis is a concerted process involving the biochemical machinery of the cell as a whole, since it consists of a series of co-ordinated reactions, the constituents of which may be synthesized in the nucleus (mRNA, tRNA), the mitochondria (ATP), the cytosol (amino acids and aminoacyl tRNAs), the ribosomes (enzymes). Although changes in the rate of protein synthesis occur, very little is known about the mechanisms involved.

Theoretically, control could be exerted at any one of the many constituent reactions, this being the pacemaker reaction for protein synthesis in that particular cell/tissue. However, the control mechanisms must allow the overall process to adjust in response to genetic, physiological and environmental stimuli, so that the rate and/or the type of proteins being synthesized can change.

2

THE CONCEPT OF ORGANELLE AUTONOMY

The most notable event in biology at the dawn of the present century was the rediscovery of the Mendelian laws of inheritance by de Vries, Correns, and Tschermak. The realization by Sutton that the results of Mendel's experiments with garden peas could be explained in terms of the visible behaviour of chromosomes during meiosis led to the nucleus being regarded as the sole carrier of the hereditary material. As often happens in science, this splendidly simple view did not survive for long. The results of experiments by Baur (1909) and Correns (1909) on the inheritance of plastid defects in variegated plants were difficult to explain on the basis that the genes concerned were located in the nucleus. In certain cases, the inheritance of such defects occurred only through the maternal line, i.e. it was uniparental.

In other cases the inheritance was biparental, but did not obey the rules of Mendel. For a time it appeared that, far from being 'nuclear', some aspects of inheritance were 'unclear'. It was Baur who pointed out that these results were explicable on the assumption of the genetic continuity of an extrachromosomal entity located in the plastid. Much more extensive evidence for this concept was later provided by the work of Renner (1929) using the genus *Oenothera;* he suggested the term 'plastome' to describe the genetic system in the plastid. Maternal inheritance is thus explicable in terms of the lack of plastids in the pollen tube of some species; the plastome of the new generation is consequently derived entirely from the mother.

These studies of variegated plants provided the first firm evidence for the existence of extrachromosomal inheritance. Since then, further evidence has accumulated with examples known in representatives of most major groups of organisms. Today we realize that organelle genetic systems are a fundamental feature of the organization of eukaryotic cells. The most detailed work has concentrated on plastids and mitochondria, but there are indications that genetic determinants may occur in other organelles found in eukaryotic cells, especially centrioles, basal granules and kinetoplasts. There is no evidence that either peroxisomes or glyoxysomes contain any genetic material. In this chapter, the biochemical evidence relating to the concept of the genetic autonomy of plastids and mitochondria only will be considered.

The concept of chloroplast autonomy was founded on the observation that, in the algae, chloroplasts can be seen to divide and to be passed to the new cells in cell division (Strasburger, 1882; Green, 1964).

This cytological evidence led to the view, first proposed by Schimper (1885) and Meyer (1893), that plastids do not arise *de novo*, but are formed by the division of pre-existing plastids; this view was supported by the genetic experiments of Baur and Correns carried out in the following two decades. The discovery in 1962 that chloroplasts contain both DNA and ribosomes brought the idea of chloroplast autonomy back into vogue, and this notion has so dominated our concepts of chloroplast development in recent years that several attempts have been made to grow isolated chloroplasts in culture. It is now clear that such attempts are ill-founded.

Our current dogmas maintain that for any biological system to be autonomous it must contain four components; (a) DNA to code for its *entire* structure; (b) DNA polymerase to replicate the DNA; (c) RNA polymerase to transcribe the DNA; (d) protein-synthesizing machinery to translate the messenger RNAs into *all* the necessary proteins. Intermediary metabolism is not necessary in principle since a supply of small molecules could be taken up from the environment. An extensive literature establishes that both chloroplasts and mitochondria do in fact contain these four components; the properties and functions of these are discussed in the rest of this chapter. It is equally clear that the DNA does not code for all the organelle proteins nor does the protein-synthesizing system make all the organelle proteins.

For example, many genes concerned with chloroplast and mitochondrial structure and function are inherited in a Medelian fashion, and are therefore presumed to be located in the nucleus, while there is increasing evidence that the majority of chloroplast and mitochondrial proteins are synthesized on cytoplasmic ribosomes. It is now realized that the demonstration of the cytological continuity of plastids from generation to generation is not sufficient to establish that they replicate independently of nuclear control. How, then, are we to regard the concept of organelle autonomy?

It is my contention that chloroplasts and mitochondria are not autonomous in any rigorous sense; the term is useful only as a quick way to describe the fact that these organelles contain *some* genes and make *some* proteins. We must regard the formation of these organelles as resulting from a complex interplay between the genomes of the organelles and the genome of the nucleus, and the fascination of this subject lies in unravelling the details of this interplay at the molecular level. Much more information is available about the autonomy of chloroplasts than about the autonomy of plant mitochondria; research on mitochondrial autonomy has concentrated almost entirely on animal and fungal cells. This imbalance is reflected in this chapter, which is concerned largely with chloroplast autonomy.

The current state of knowledge can be summarized by saying that, while some information has been accumulated about the properties of chloroplast nucleic acids and protein synthesis, only recently has any hard evidence emerged about their biological function. Developments at the genetic level in higher plants, and the establishment of isolated chloroplast systems which synthesize specific protein and RNA molecules, promise to provide increasing understanding of the precise functions of chloroplast nucleic acids and protein synthesis.

CHLOROPLAST AUTONOMY

The existence of DNA in chloroplasts has been established by both cytological and biochemical criteria. Ris and Plaut (1962) provided the first convincing evidence that chloroplasts contain DNA. Electron microscopy of chloroplasts in sections of both algal and higher plant cells revealed areas of low electron density which contained fibrils 2.5–3.0 nm in diameter. These fibrils were not seen if the sections were treated with deoxyribonuclease. There are earlier reports that chloroplasts contain regions which can be stained with the Feulgen reagent for DNA, but in general these reports have not proved reproducible.

It is now clear that in most species the concentration of DNA inside the chloroplast is below the limit of detection by the Feulgen method. It can however be detected by autoradiography after exposure of cells to -H-thymidine; the report of Rose *et al.* (1974) on the distribution of DNA in dividing spinach chloroplasts contains an excellent example of this method. It is important when evaluating such studies to consider the controls that are used. If the incorporated label is all present in DNA, it should be removed by treatment of sections with deoxyribonuclease, but not by treatment with ribonuclease. This control works well for tobacco and spinach leaves and several algae, but not or maize leaves, where deoxyribonuclease fails to remove the label from chloroplasts.

Most studies on chloroplast DNA since 1962 have been carried out using isolated chloroplasts. This approach suffers from the difficulty of distinguishing chloroplast DNA from DNA originating from nuclei, mitochondria, and microorganisms which may contaminate the chloroplast pellet. The resolution of these problems is still not complete, and depends mainly on the differing properties of the DNA from the various sources.

There are four characteristic properties of chloroplast DNA that have been used to identify it: namely buoyant density, ease of renaturation, lack of histones, and the absence of 5-methylcytosine. These properties are discussed below. It must be remembered that there is a possibility some chloroplast DNA may not meet these criteria. If, for example, chloroplasts in some species contain a minor DNA component which is a nuclear transcript, contains 5-methylcytosine and renatures poorly, it is doubtful whether present techniques could identify it as chloroplast in location.

In one species the problem of contamination by nuclei and micro-organisms can be entirely avoided; DNA has been identified in chloroplasts isolated from sterile enucleated plants of *Acetabularia* (Gibor & Izawa, 1963; Baltus & Brachet, 1963).

Buoyant Density

It is now clear that chloroplast DNA from the algae *Euglena, Chlamydomonas,* and *Chlorella* has a buoyant density sufficiently different from that of nuclear DNA to allow resolution by analytical ultracentrifugation, but that in many higher plants the densities of the two DNA types are often, but not always, too close to permit this.

This information has been hard-won; the story of the way in which the interpretation of band patterns in neutral caesium chloride density gradients has changed since 1963 has been told by Kirk (1971) and Tewari (1971). A summary is given here since it illustrates the very real problem of establishing the identity of chloroplast DNA from higher plants by biochemical methods.

The first report of the chemical characterization of higher plant chloroplast DNA was provided by Kirk (1963). He found that chloroplast DNA from *Vicia faba* has a GC content (37.4 %) which is slightly but significantly different from that of nuclear DNA (39.4%). In the same year a contrasting report was published by Chun *et al.,* (1963), who found that chloroplast preparations from *Spinacia oleracea* and *Beta vulgaris* contained two components of much higher densities than the nuclear DNA as judged by caesium chloride equilibrium density gradient centrifugation.

However the bulk of the DNA in these chloroplast preparations had a density similar to that of nuclear DNA; this band was attributed to contamination of the chloroplast pellet by nuclear fragments, and thus the chloroplast DNA was regarded as the two high density components. There followed a spate of papers which supported the claim that, in higher plants, chloroplast DNA has an appreciably higher density than nuclear DNA.

This picture changed when re-examination by more rigorous methods supported the earlier view of Kirk, and the higher density components are now attributed to contamination by mitochondria and bacteria.

Table . Buoyant densities and base composition of higher-plant and algal chloroplast and nuclear DNA*.

	Nuclear DNA		Chloroplast DNA	
Species	Density in CsCl (g cm^{-3})	%GC** content	Density in CsCl (g cm^{-3})	% GC content
Chlorella ellipsoidea	1.716	60	1.692	33.0
Euglena gracilis	1.707	52	1.685	25.0
Chlamydomonas reinhardi	1.723	67	1.695	39.0
Vicia faba	1.695	38.8	1.697	37.8
Vicia faba	1.696	39.8	1.696	36.8
Spinacia oleracea	1.694	37.8	1.697	37.8
Lactuca sativa	1.694	37.8	1.697	37.8
Pisum sativum	1.695	38.8	1.697	37.8
Nicotiana tabacum	1 697	40.8	1.697	37.8
Nicotiana tabacum	1.697	40.8	1.700	40.8
Nicotiana tabacum	1.695	38.8	1.697	38.0
Phaseolus vulgaris	1.697	40 8	1.696	36.7
Phaseolus vulgaris	1.696–1.697	40.0	1.696–1.697	37–38
Allium cepa	1.691	34.7	1.696	37.2
Triticum vulgare	1.702	47.0	1.697	38.4
Oenothera hookeri	1.703	47.0	1.697	38.0
Antirrhinum majus	1.691	34.7	1.697	38.0

* This table is taken partly from a compilation by Kirk (1971).

** In order to correct for the lowering of the buoyant density of the nuclear DNAs caused by methylation of cytosine residues, a value of 0.003 was added to all the observed densities of nuclear DNA before calculating the GC content.

(Reproduced with permission from MTP International Review of Science, Biochemistry Series One, Vol. 6, *'Biochemistry of Nucleic Acids'*, 1974, ed. by K. Burton, and published by Butterworths, London.)

The present position can be summarized by saying that, in all the higher plants examined, chloroplast DNA had a base composition of 37.5 ± 1% GC and a buoyant density of 1.697 ± 0.001 g cm^{-3}. Table 11.1 lists some values for the buoyant densities of chloroplast and nuclear DNA which have been agreed by Kirk and Tewari. Nuclear DNA is seen to have a smaller, larger, or similar density to chloroplast DNA from the same plant, depending on the species.

The GC contents in Table below have been calculated from the buoyant densities, but a more accurate method has been devised by Kirk (1967). This method releases the purine bases by gentle acid hydrolysis and separates the adenine and guanine by ion-exchange chromatography.

The moral to be drawn from this story is that the purity of subcellular fractions from higher plant tissues must be established by positive methods if meaningful interpretations are to be made. Microbial contamination can be reduced by surface sterilization of fresh tissues and by the use of sterile solutions, while mitochondrial contamination can be assayed by succinic dehydrogenase or cytochrome oxidase activities. Nuclear contamination can be reduced by incubation of intact chloroplasts with deoxyribonuclease and phosphodiesterase. The extent of nuclear contamination is difficult to measure, but the ease of renaturation and the absence of 5-methylcytosine serve to distinguish chloroplast from nuclear DNA in all cases so far studied.

Ease of Renaturation

A more useful criterion than the buoyant density for detecting chloroplast DNA from higher plants is the ease with which it will renature after heat or alkali denaturation. Nuclear DNA renatures only to a slight extent in a few hours, depending on its content of reiterated sequences.

The rapid renaturation of chloroplast DNA has encouraged attempts to estimate its genome size from measurements of kinetic complexity. The kinetic complexity of a DNA sample is a measure of the size of the unique set of nucleotide sequences it contains, as judged from the rate at which the DNA renatures.

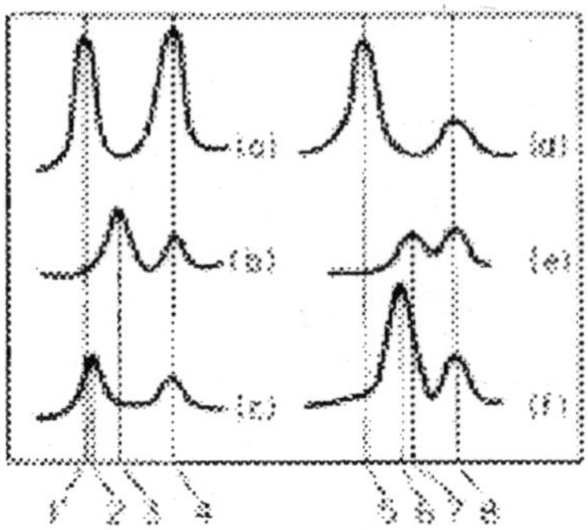

Figure: Densitometer tracings of ultraviolet absorption photographs of* Vicia faba *DNA banded in caesium chloride gradients. (a) Native chloroplast DNA; (b) heat-denatured chloroplast DNA; (c) renatured chloroplast DNA; (d) native nuclear DNA; (e) heat-denatured nuclear DNA; (f) renatured nuclear DNA. The numbers refer to densities in g cm^{-3} as follows: (1) 1.696; (2) 1.699; (3) 1.712; (4) 1.728; (5) 1.696; (6) 1.709; (7) 1.712; (8) 1.728. The peaks at 1.728 g cm^{-3} represent marker DNA from* Micrococcus radiodurans. *(From Kung & Williams, 1968, by courtesy of Elsevier.)

A rapid rate of renaturation implies that like sequences are present in high concentration, and therefore the number of unique sequences, or kinetic complexity, is small. Table below lists the kinetic complexity, in terms of molecular weight, for the chloroplast DNA from some algae and higher plant species.

Some of these values have been corrected from the published figures since the estimate of the molecular weight of the bacteriophage T4 DNA, used as standard, has been revised from 1.3×10^8 to 1.06×10^8 (Dubin *et al.*, 1970). It is striking that the corrected values for the kinetic complexity of chloroplast DNA from the few algae and higher plants examined so far are all in the range $0.9–1.0 \times 10^8$.

This may be a coincidence, or it could mean the information content of chloroplast DNA is basically similar throughout the plant kingdom. All the species so far examined belong to the Chlorophyta, with the exception of *Euglena*. It would be interesting to make such measurements of chloroplast DNA from other plant groups, especially from algae with unusually shaped chloroplasts. Table below also lists the analytical complexities of chloroplast DNA, i.e. the amount of DNA per chloroplast. Since the kinetic complexities are always much less than the analytical complexities, there must be between 20 and 60 copies of the DNA sequences in each chloroplast.

These renaturation studies cannot rule out the possibility of microheterogeneity in nucleotide sequence between the copies but, if it does exist, such heterogeneity is beyond detection by current techniques.

Table . Corrected kinetic and analytical complexities of chloroplast DNA.

Species	Kinetic Complexity (Daltons × 10^8)	Analytical Complexity (Daltons per chloroplast × 10^9)	Reference
Euglena gracilis	0.9	5.8	Stutz (1970)
Chlamydomonas			Wells & Sager (1971);
reinhardi (gamete)	0.99	4–5	Bastia *et al.*, (1971)
Pisum sativum	0.95	—	Kolodner & Tewari (1972)
Lactuca sativa	0.98	2	Wells & Birnstiel (1969)
Nicotiana tabacum	0.93	3	Tewari & Wildman (1969)

(Reproduced with permission from MTP International Review of Science, Biochemistry Series One, Vol. 6. *'Biochemistry of Nucleic Acids'*, 1974. Edited by K. Burton and published by Butterworths, London.)

Absence of Histones and 5-Methylcytosine

When released from intact isolated chloroplasts by gentle lysis, chloroplast DNA is not combined with basic proteins. This finding confirms the microscopic observations of Ris and Plaut (1962) that chloroplasts contain DNA fibrils in the same form as they appear in the nucleoplasms of prokaryotic cells. It is also characteristic of chloroplast DNA from both higher plants and algae that it contains no detectable 5-methylcytosine, whereas nuclear DNA invariably does. Whitfeld and Spencer (1968) regard the absence of this base as the most reliable criterion for establishing the purity of chloroplast DNA.

Circularity

A recent discovery is that, when precautions are taken to minimize shearing, a proportion of chloroplast DNA can be isolated as circles. Circular DNA has been reported from chloroplasts of *Euglena* (Manning *et al.*, 1971), *Pisum sativum* (Kolodner & Tewari, 1972), *Spinacia oleracea* (Manning *et al.*, 1972), *Antirrhinum majus*, *Oenothera hookeri*, and *Beta vulgaris* (Hermann *et al.*, 1975). The significance of circularity in DNA is not known, but it is a useful property since it implies the molecule has not been degraded on isolation. The most interesting aspect of this finding is that in all the species listed the contour length of the majority of the circles is in the range 37–45 μm; this length of double-stranded DNA has a calculated molecular weight of $0.85–1.0 \times 10^8$ daltons, which is in the same range as the kinetic complexity of chloroplast DNA. This correspondence of length and kinetic complexity suggests that the genetic information carried by the chloroplast DNA is accommodated by the length of the circular molecule.

Ploidy

It is clear from the data in Table below, and from the evidence for circular DNA, that each chloroplast may contain many copies of a circular genome, and thus may be genetically polyploid. It has been pointed out by Kirk (1972) that this multiplicity increases the probability that a mutation will appear in a chloroplast in a given time, but there is likely to be a long delay before the mutation can be expressed, since a mutation in one copy will be swamped by all the other wild-type copies. It is therefore not surprising that known non-Mendelian mutations affecting chloroplasts are small in number, and difficult to induce with mutagens. In *Chlamydomonas reinhardi* a number of mutations are known which alter the sensitivity of chloroplast ribosomes to inhibition by antibiotics such as erythromycin and carbomycin; some of these mutations show uniparental inheritance (Sager, 1972; Mets &

Bogorad, 1972). Those genes which are inherited in a uniparental fashion have been shown by recombinational analysis to form one linkage group which behaves as if it were diploid in vegetative cells. It is difficult to reconcile this diploid behaviour with the evidence which suggests that there are many copies of the genome in each chloroplast. It may be that the uniparental linkage group does not reside in chloroplast DNA, or if it does, that a reduction in the number of genomes in each chloroplast occurs at some stage in the cell cycle, so that only a 'master copy' is passed during sexual reproduction. There is as yet no evidence to resolve these questions.

Functions

A 40 µm circle of double-stranded DNA of unique base sequence is sufficient in principle to code for about 125 proteins each of molecular weight 50,000. Since our knowledge depends ultimately on the validity of the techniques which can be used, the evidence for the functions of chloroplast DNA will be considered in terms of the four methods tried so far:

The Selective Inhibition of Chloroplast DNA Transcription

The antibiotic rifampicin is a potent inhibitor of the initiation of RNA synthesis in bacteria. This compound has been reported to inhibit the incorporation of labelled precursors into chloroplast ribosomal-RNA, but not into cytoplasmic ribosomal-RNA, in several unicellular algae, namely *Chlamydomonas, Chlorella,* and *Acetabularia.* Rifampicin has also been reported to inhibit chloroplast RNA polymerase activity in extracts of *Chlamydomonas.* These results suggest that the functional genes for chloroplast ribosomal-RNA are located in chloroplast DNA and are transcribed by the chloroplast RNA polymerase.

The effect of rifampicin in higher plant systems is controversial. There have been some reports that it specifically inhibits chloroplast ribosomal-RNA synthesis, but other workers could not reproduce this result (Bottomley *et al.,* 1971). There now seems general agreement that chloroplast RNA polymerase from higher plants, as normally prepared and assayed, is insensitive to rifampicin, but this may mean only that the preparations are not initiating RNA synthesis.

Genetic Analysis of Mutants

Many mutations are known which affect chloroplast components, but the vast majority are inherited in a Mendelian fashion and are therefore presumed, to be located in nuclear genes. For example, seven different genes are known which control steps in the chlorophyll biosynthetic pathway, and nuclear genes have been shown to be involved in the synthesis of phosphoribulokinase and at least five components of the photosynthetic electron transport chain (Kirk, 1972; Kirk & Tilney-Bassett, 1967; Levine & Goodenough, 1970).

In *Chlamydomonas* many mutations affecting chloroplast ribosomes and photosynthetic capacity are inherited as a single linkage group in a uniparental fashion (Sager, 1972; Schlanger & Sager, 1974).

In this alga, like gametes fuse completely, and the physical basis for uniparental inheritance has been suggested to be the destruction of the chloroplast DNA of one of the gametes by a modification-restriction system of the type known in bacteria. As pointed out earlier, there are difficulties in concluding that this uniparental class of mutations resides in chloroplast DNA, but it is tempting to believe that it does. A simple interpretation suggests that some of the proteins of the chloroplast ribosomes are encoded in chloroplast DNA; there is also evidence that at least one other protein of the chloroplast ribosome is encoded in a nuclear gene (Mets & Bogorad, 1972).

Rigorous proof that a mutation inherited in a maternal or uniparental fashion actually alters the amino acid sequence of a protein has been obtained in only one case. Chan and Wildman (1972) studied the inheritance of a mutation in the large subunit of Fraction I protein in *Nicotiana tabacum* at the tryptic peptide level; this mutation was inherited *via* the maternal line only. By contrast, mutations in the small subunit of Fraction I protein are inherited in a Mendelian fashion (Kawashima & Wildman, 1972).

This type of combined biochemical-genetic approach is very promising and deserves further use, especially with cultivated plants where many varieties are available. It must be emphasized that this approach must be conducted at the level of the tryptic peptide analysis of a purified protein. It is not sufficient to show that particular proteins are absent in a mutant variety, since absence of a protein might result from a mutation in a chloroplast gene which controls the formation of a protein encoded in the nucleus.

DNA-RNA HYBRIDIZATION STUDIES

If RNA isolated from chloroplasts can be shown to hybridize to chloroplast DNA but not to nuclear DNA, this is good grounds for believing that such RNA is both encoded in and transcribed from chloroplast DNA. A number of hybridization studies have been carried out with chloroplast ribosomal-RNA, which hybridizes to 0.5–6.0% of chloroplast DNA (Ellis & Hartley, 1974). The most recent study is that of Thomas and Tewari (1974); they found that in a number of higher plants each circle of chloroplast DNA contains two genes for chloroplast ribosomal-RNA.

There are several reports that chloroplast ribosomal-RNA will also hybridize to nuclear DNA. The significance of these reports awaits further study, but the possibility is raised that there are two types of chloroplast ribosomal-RNA, one encoded in chloroplast DNA and the other in nuclear DNA. This arrangement could be a means whereby the nucleus exerts control over events in the chloroplasts, especially if it is further postulated that chloroplast ribosomes containing nuclear RNA translate only messengers originating in the nucleus.

There is evidence that some of the transfer RNA species found in chloroplasts will also hybridize to chloroplast DNA. Tewari and Wildman (1970) found that tRNA from tobacco chloroplasts would hybridize to 0.4–0.7% of chloroplast DNA. This amount of DNA would be enough to code for 20 to 30 transfer RNA molecules, each of molecular weight 25,000. This work used unfractionated tRNA however, and more studies need to be done to establish the site of encoding of individual tRNA species.

Identification of RNA and Protein Molecules Synthesized by Isolated Chloroplasts

This is the most direct method; if transcription coupled to translation can be obtained in isolated chloroplasts, identification of the products would simultaneously determine both the structural genes present in chloroplast DNA and the function of chloroplast ribosomes.

Work in the author's laboratory has shown that isolated intact chloroplasts from *Pisum sativum* and *Spinacia oleracea* will synthesize discrete protein and RNA molecules, but transcription and translation are not coupled (Hartley & Ellis, 1973; Ellis, 1974; Ellis & Hartley, 1974). It is therefore not possible to infer that the proteins synthesized by isolated chloroplasts are encoded in chloroplast DNA.

Figure: Analysis by polyacrylamide gel electrophoresis of RNA synthesized by chloroplasts isolated from **Spinacia oleracea. *Isolated chloroplasts were incubated with ^{3} H-uridine and 15,000 lux of red light for 20 min at 20°C. Nucleic acid was extracted and run on polyacrylamide gels. The solid line represents the A_{260}, while the histogram shows the radioactivity. The figures are the molecular weights × 10^{-6} of the RNA components. (Reprinted from Hartley & Ellis, 1973, by courtesy of the Biochemical Journal).***

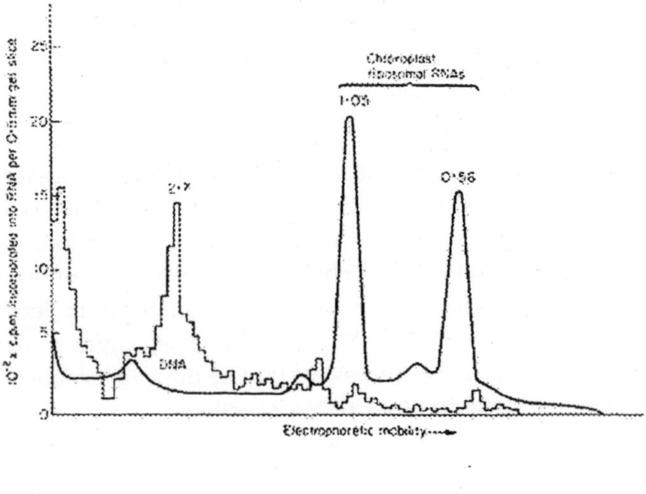

The RNA synthesized by isolated chloroplasts has been analyzed by polyacrylamide gel electrophoresis. The chief product is a species of molecular weight about 2.7×10^6 ; this has been shown by competitive hybridization to chloroplast DNA to be a precursor to chloroplast ribosomal-RNA. Isolated chloroplasts have not been shown convincingly to synthesize ribosomal-RNA; presumably the processing system which trims the precursor molecules does not survive the trauma of isolation.

The known genes in chloroplast DNA can be summarized as follows: chloroplast transfer-RNA, chloroplast ribosomal-RNA, several chloroplast ribosomal proteins, and the large subunit of Fraction I protein. These genes account for less than 10% of the total potential coding capacity of a 40 µm circle of DNA. The elucidation of the function of the remaining 90% of chloroplast DNA is the most important problem in this field.

Chloroplast DNA Polymerase

DNA polymerase activity has been detected in chloroplast suspensions prepared from *Spinacia oleracea. Nicotiana tabacum,* and *Euglena gracilis.*

The product renatures readily after heat denaturation, and hybridizes to a much larger extent with chloroplast DNA than with nuclear DNA. It has not been established whether the chloroplast polymerase is identical with any nuclear polymerase or whether it carries out either a replicase or a repair function.

The enzyme is bound to the chloroplast membranes, but in the case of *Euglena,* it has been solubilized by treatment with high concentrations of salt and highly purified; the purified enzyme is inhibited by ethidium bromide (Keller *et al.,* 1973). Flechtner and Sager (1973) have made the interesting observation that treatment of *Chlamydomonas* cells with ethidium bromide induces a selective and reversible inhibition of chloroplast DNA replication.

Thus nuclear DNA synthesis proceeds normally while replication of chloroplast DNA is impaired; the preexisting chloroplast DNA decreases by at least 80% in one cell generation, but this loss is reversible if the drug is removed within 12 hours.

This result suggests that one or a few of the chloroplast DNA copies may be protected in some way, perhaps by close attachment to chloroplast membranes. Such sequestered DNA could act as a 'master' copy, and might account for the diploid behaviour of the uniparental linkage group in this organism. The chloroplast DNA of *Euglena* and *Chlamydomonas* has been shown to replicate in a semi-conservative fashion at a different time in the growth cycle from nuclear DNA, but the factors controlling the time and rate of synthesis of chloroplast DNA are unknown.

Chloroplast RNA Polymerase

DNA-dependent RNA polymerase activity has been demonstrated in chloroplast preparations from several species of algae and higher plants. Most workers study lysed chloroplast preparations because the nucleotide triphosphates used as substrates penetrate the chloroplast envelope at slow rates.

Unlike intact chloroplasts, which use light to phosphorylate added ^{3}H-uridine, lysed preparations do not synthesize discrete species of RNA; instead a polydisperse pattern of products ranging in size from 5s to 23s is obtained (Spencer & Whitfeld, 1967). This difference between the products from intact and lysed chloroplasts may result from the dilution of controlling factors on lysis.

The RNA polymerase activity of chloroplast preparations can represent a high proportion of the total RNA polymerase activity measurable in leaf extracts.

There are two key questions about chloroplast RNA polymerase which need to be answered: firstly, how similar is it to any nuclear polymerase, and secondly, how is its activity regulated ? The enzyme is bound to the chloroplast lamellae, but a technique for its quantitative removal has been devised (Bennett & Ellis, 1973).

The solubilized chloroplast polymerase has been purified, and appears very similar to a nuclear polymerase (Bogorad *et al.*, 1973), but a detailed characterization has yet to be reported. It has been shown that the synthesis of chloroplast ribosomal-RNA in dark-grown plants is phytochrome-mediated (Scott *et al.*, 1971), but the nature of the mechanism which links phytochrome to RNA polymerase is unknown.

Figure: Distribution of chloroplast RNA polymerase activity in leaves of young shoots of **Pisum sativum.** ***Pea seedlings were grown for 14 days and the leaves homogenized. Total RNA polymerase activity was measured in the low-speed pellet containing nuclei and chloroplasts, and chloroplast RNA polymerase activity in purified chloroplast membranes. The figures are the activities of the chloroplast RNA polymerase as percentages of the total activity. (Unpublished work of J. Bennett.)***

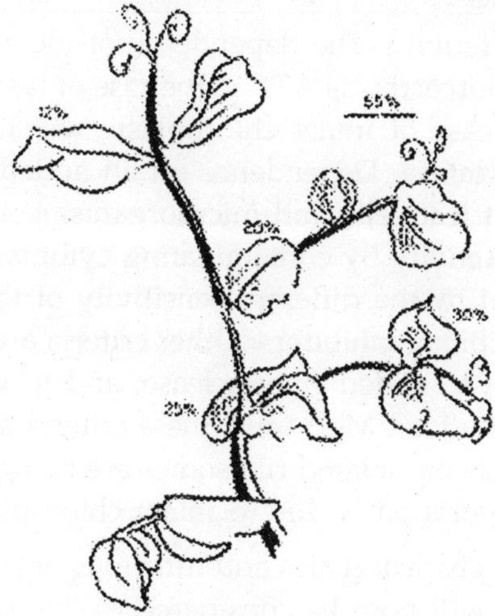

Chloroplast Protein Synthesis

Chloroplast preparations capable of incorporating labelled amino acids into protein have been isolated from higher plants and algae. Table below illustrates some results obtained with preparations from young bean and tobacco leaves; preparations from mature leaves have little activity.

These results are typical of those reported in many studies with many species (Ellis *et al.*, 1973). In such work it is vital to establish that the incorporation is due to chloroplasts and not to microorganisms, intact leaf cells, or cytoplasmic ribosomes.

Table . Characteristics of [^{14} C]-leucine incorporation into protein by isolated lysed chloroplasts. A pellet containing chloroplasts was resuspended in sterile 50 mM Tris-HCl, 10 mM $MgCl_2$, 100 mM KCl, 5 mM 2-mercaptoethanol (pH 7.8), and incubated for 40 min at 25°C with 2 mM ATP, 0.2 mM GTP, and 1 μC [^{14} C]-leucine. Total protein was extracted and counted. The incorporation obtained in the complete system is called 100. (Unpublished results of R.J. Ellis.)

Assay Conditions	*Phaseolus vulgaris*	Nicotiana tabacum
Complete	100	100
Zero time	0.5	1.5
GTP and ATP omitted	3	5
Ribonuclease added	5	6
Puromycin added	21	22
Chloramphenicol added	23	25
Actinomycin D added	100	100
Cycloheximide added	90	95

The best criterion is the dependence of incorporation on an added energy source; this is ATP in the case of lysed chloroplasts, or light in the case of intact chloroplasts which can carry out photophosphorylation. Dependence on an added energy source eliminates intact leaf cells and microorganisms as the agents of incorporation. Activity by contaminating cytoplasmic ribosomes can be ruled out by the different sensitivity of the two types of ribosome to antibiotic inhibitors. Other criteria are the sensitivity of incorporation to added ribonuclease, and to variation in the concentration of added Mg^{2+} ions; these criteria are useful when lysed chloroplasts or isolated ribosomes are being tested, but not when the incorporation is due to intact chloroplasts.

Some of the characteristics and functions of protein synthesis by chloroplasts will now be considered.

Ribosomes

Lyttleton (1962) was the first to show that green plant cells contain two classes of ribosome which differ in their sedimentation coefficients. Chloroplasts contain 70s ribosomes while the cytoplasm contains 80s ribosomes. Figure below shows a density gradient analysis of an homogenate of spinach leaves, from which it can be seen that chloroplast ribosomes constitute a high proportion of the total cellular complement of ribosomes; values as high as 60% have been reported.

Figure: Proportion of chloroplast ribosomes in leaves of **Spinacia oleracea.** ***Extracts of spinach leaves were treated with detergent to release membrane-bound ribosomes, and all the ribosomes collected by centrifugation. The ribosomes were analysed on a linear sucrose density gradient. (Unpublished work of M. R. Hartley.)***

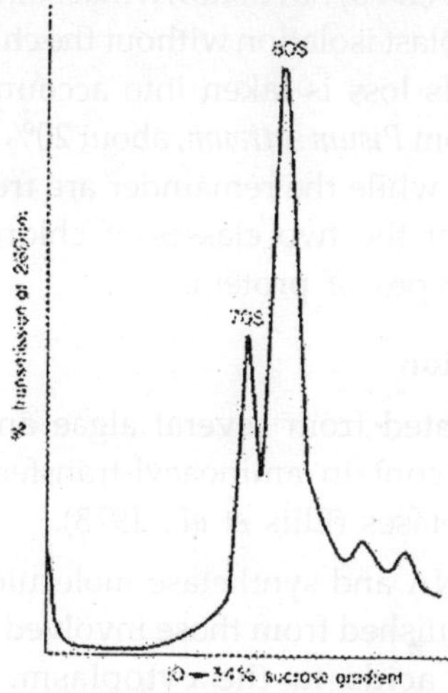

Chloroplast ribosomes resemble those from prokaryote cells in their s value. Their RNA components are also of the same size as those found in prokaryote ribosomes, i.e. 16s and 23s. The 80s cytoplasmic ribosomes by contrast, contain 18s and 25s RNA molecules. These similarities between chloroplast and prokaryote ribosomes have been much stressed in the past, but it is now clear that these similarities are of size only, and not in the primary structure of the RNA and protein components. For example, ribosomal-RNA from *Escherichia coli* does not compete with chloroplast ribosomal-RNA from *Euglena* for hybridization to chloroplast DNA. The protein complement of prokaryote ribosomes is also quite different from that of chloroplast ribosomes as judged by gel electrophoresis and immunological tests. Similarly, the proteins of chloroplast and cytoplasmic ribosomes from the same plant differ distinctly in the patterns they give on gel electrophoresis.

Some of the chloroplast ribosomes are bound to the internal membranes. Electron microscopy has revealed whorl-like polyribosomes attached to the granal and intergranal thylakoid membranes, while analysis of isolated chloroplasts shows that up to 50% of the ribosomes cannot be removed from the membranes by washing with hypotonic buffer.

Tao and Jagendorf (1973) have shown that some free ribosomes are lost during chloroplast isolation without the chloroplasts lysing irreversibly; when this loss is taken into account, they estimate that in chloroplasts from *Pisum sativum*, about 20% of the ribosomes are membrane-bound while the remainder are free in the stroma. There is evidence that the two classes of chloroplast ribosome synthesize different types of protein.

Amino Acid Activation

Chloroplasts isolated from several algae and higher plants have been found to contain aminoacyl-transfer-RNAs and the corresponding synthetases (Ellis *et al.*, 1973).

These transfer RNA and synthetase molecules can often, but not always, be distinguished from those involved in the activation of the same amino acids in the cytoplasm. Studies of the compatibility between transfer-RNA and synthetases from the two cellular compartments show that the specificity of interaction ranges from. none to absolute.

For example, the cytoplasm of leaves of *Phaseolus vulgaris* contains a leucyl-transfer-RNA synthetase which can add leucine only to the two leucyl-transfer-RNA species common to the cytoplasm and chloroplasts, but is not able to recognize the three leucyl-transfer-RNA species found only in the chloroplasts. The chloroplast enzyme, on the other hand, can aminoacylate all five leucyl-transfer-RNA species.

It is difficult to attach any biological significance to these variations in specificity. The simplest presumption is that in the intact cell the chloroplast synthetases aminoacylate only the transfer-RNA species located in the chloroplast and these are sufficient for chloroplast protein synthesis to proceed, while the cytoplasmic synthetases aminoacylate only the transfer-RNA species located in the cytoplasm.

A more interesting possibility is that some of the transfer RNA species required to translate chloroplast messenger-RNA are acylated in the cytoplasm; such a requirement would provide one way in which protein synthesis in the cytoplasm and chloroplast could be integrated, but there is no direct evidence to support this suggestion. Experiments with *Euglena* suggest that at least some of the chloroplast aminoacyl-transfer-RNA synthetases are encoded in nuclear DNA and are synthesized on cytoplasmic ribosomes (Hecker *et al.*, 1974).

Initiation of Chloroplast Protein Synthesis

Protein synthesis in bacteria is initiated by N-formyl-methionyl transfer RNA; there is evidence that this is also true in chloroplasts, but not in the cytoplasm of eukaryotic cells.

Schwarz *et al.*, (1967) found that chloroplast ribosomes from *Euglena* would translate RNA from bacteriophage f2 into viral coat protein with N-formylmethionine at the amino terminus. Detailed studies of initiating methionyl-transfer-RNA species have since been reported for several higher plants.

For example, there are two methionyl-transfer-RNA species in wheat chloroplasts. One of these can be formylated by a chloroplast transformylase, whereas the other cannot, and may serve to direct methionine into internal positions in the polypeptide chains (Leis & Keller, 1971).

Two other methionyl-transfer-RNA species are found in the cytoplasmic fraction, neither of which can be formylated. It is probable from this type of evidence that the initiation of protein synthesis by chloroplasts is similar to that in prokaryotes in that it uses a formylated methionyl-transfer-RNA, but distinct from that in the cytoplasm which uses an unformylated methionyl-transfer-RNA.

The significance of this distinction is not clear; it may reflect the evolutionary origin of the compartments in eukaryotic cells, and have no contemporary functional value.

Energy Source

In most studies of protein synthesis by isolated chloroplasts, added ATP has been used as an energy source. Since isolated chloroplasts can generate ATP by photophosphorylation, it should be possible to drive protein synthesis in chloroplasts with light. Spencer (1965) found that spinach chloroplasts would use light to incorporate amino acids into protein provided that ADP, inorganic phosphate and pyocyanine were added.

The necessity for pyocyanine as catalyst indicates that the chloroplasts were broken, and had lost their natural catalyst, ferredoxin, by dilution. If precautions are taken to isolate chloroplasts which have their outer envelopes intact, protein synthesis is stimulated twenty-fold by light in the absence of either cofactors or catalysts of photophosphorylation.

Figure: Light-driven protein synthesis by chloroplasts isolated from* Pisum sativum. *Intact isolated chloroplasts were incubated with ^{14}C-leucine and 4,000 lux red light at 20°C. Total protein was then extracted and its radioactivity measured. (Reprinted from Blair & Ellis, 1973, by courtesy of Elsevier.)

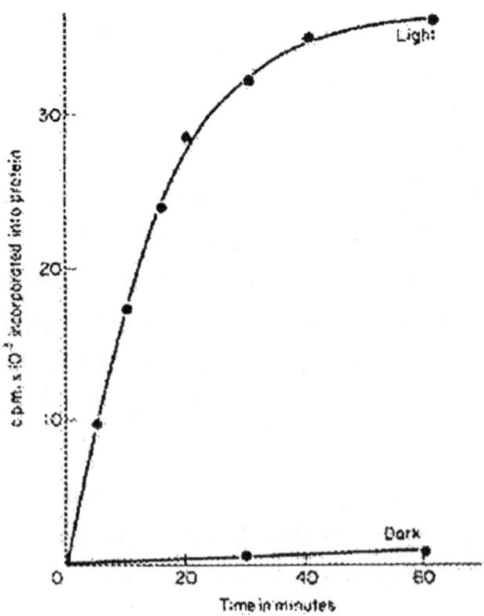

The rates of light-driven protein synthesis by such chloroplasts are the highest yet recorded for any isolated chloroplast system. The use of intact chloroplasts for studies of protein synthesis has a major advantage over the use of broken chloroplasts in that conditions around the polysomes are more normal with respect to controlling factors.

It is thus more likely that correct termination and release of polypeptide chains occurs in intact chloroplasts than in lysed preparations; such intact chloroplasts make discrete protein molecules rather than incomplete polypeptide chains. A further advantage of using light as the energy source is that the broken chloroplasts, which inevitably contaminate the preparation of intact chloroplasts, cannot contribute to the incorporation in the absence of added cofactors.

It is possible that protein synthesis by chloroplasts *in vivo* uses ATP provided by the chloroplast itself.

However, this is not the case in developing chloroplasts, since organisms such as *Chlorella, Chlamydomonas,* and *Pinus* do not require light to form chloroplasts. The formation of chloroplasts must therefore depend on ATP supplied by the rest of the cell. This conclusion is confirmed by the ability of intact isolated etioplasts from *Pisum sativum* to use ATP, but not light, as their source of energy for protein synthesis (Siddell & Ellis, 1975).

Inhibitors

It is well established that protein synthesis by isolated chloroplast ribosomes is inhibited by the same antibiotics which inhibit protein synthesis by prokaryote ribosomes. The best studied example is chloramphenicol, which inhibits protein synthesis by isolated chloroplasts from all the species so far tested; inhibition is shown only by the D-threo isomer of chloramphenicol, and not by the other stereo isomers.

Three other unrelated antibiotics, spectinomycin, lincomycin, and erythromycin also inhibit protein synthesis by chloroplast ribosomes. Light-driven protein synthesis by chloroplasts from *Pisum sativum* is especially sensitive to lincomycin, 50% inhibition being given by 0.2 μg ml^{-1} (Ellis & Hartley, 1971). Protein synthesis by cytoplasmic ribosomes from plants is not inhibited by any of these antibiotics.

On the other hand, cytoplasmic ribosomes are inhibited by cycloheximide, which does not affect the activity of chloroplast ribosomes. It must be emphasized that all these findings relate only to the activity of isolated sub-cellular systems; it cannot be assumed from these data alone that it is valid to use these compounds on intact cells with the expectation of inhibiting protein synthesis in one compartment but not in the other.

The protein complements of chloroplast and bacterial ribosomes are different, and thus the above similarity between them must reside in their antibioticbinding sites, and not in the properties of the proteins as revealed by gel electrophoresis or immunological tests.

There is no evidence that the mechanism of action of bacterial antibiotics on chloroplast ribosomes is similar to that on bacterial ribosomes; nor is there any understanding of the selective pressure which maintains the sensitivity of chloroplast ribosomes to these antibiotics in the face of mutations to resistance, which can, for example, be found in *Chlamydomonas*.

Functions

The high proportion of plant ribosomes located inside chloroplasts raises the question as to their function. Are they required in such quantities because they make a large number of different chloroplast proteins, or because they make a small number of different chloroplast proteins in large amounts? The available evidence suggests that the latter is the case (Ellis *et al.*, 1973; Ellis, 1974).

The problem of identifying which proteins are synthesized by chloroplast ribosomes has been tackled in two ways. The first is to supply antibiotic inhibitors, especially chloramphenicol and cycloheximide, to intact cells making chloroplasts, and determining which proteins are no longer synthesized. The validity of results from experiments with any inhibitor depends absolutely on the specificity of its action in intact cells, and there is evidence that both chloramphenicol and cycloheximide have effects on activities other than protein synthesis in some higher plants (Ellis & MacDonald, 1970).

Processes such as ion uptake, oxidative phosphorylation and photophosphorylation are inhibited by all four stereoisomers of chloramphenicol, whereas the inhibition of protein synthesis by isolated chloroplast ribosomes is specific for the D -threo isomer (Ellis, 1969).

This stereospecificity provides a means of establishing for any particular tissue whether chloramphenicol is inhibiting protein synthesis directly at the ribosomal level, or in addition, is affecting some other process such as energy supply. It is strongly recommended that only if an inhibition is produced specifically by the D -threo isomer should an interpretation directly involving protein synthesis be invoked.

Another problem is that in most of the inhibitor experiments on chloroplast ribosomal function, increases in specific proteins have been measured as enzymic activities rather than as amounts of protein. Failure to observe an effect by a particular inhibitor might therefore mean that the increase in enzymic activity is due to activation of a precursor protein, rather than to *de novo* synthesis by either chloroplast or cytoplasmic ribosomes.

Bearing these difficulties in mind, this author interprets the bulk of the published inhibitor experiments as suggesting that most of the chloroplast proteins are synthesized, on cytoplasmic ribosomes (Ellis *et al.*, 1973; Ellis, 1975).

In all the studies reported on several algae and higher plants, the synthesis of Fraction I protein was found to be inhibited by 70s ribosomal inhibitors. In most cases, the synthesis of the other soluble enzymes of the photosynthetic carbon dioxide reduction cycle appears to occur on cytoplasmic ribosomes; the same is true for ferredoxin and the chloroplast RNA polymerase.

Besides Fraction I protein, the only other proteins which appear to be synthesized by chloroplast ribosomes are some of the chloroplast ribosomal and lamellar proteins, including the photosynthetic cytochromes. It must be emphasized, however, that these inhibitor experiments are never more than suggestive. Strictly interpreted, they never say more than that the activity of a particular group of ribosomes is required for a particular protein to accumulate in the chloroplast.

This is not the same as saying that these ribosomes actually synthesize that protein, because it is possible that the apoenzyme is synthesized by one class of ribosomes but requires for its appearance in the chloroplast in an active state, additional protein(s) which are synthesized by another class of ribosomes.

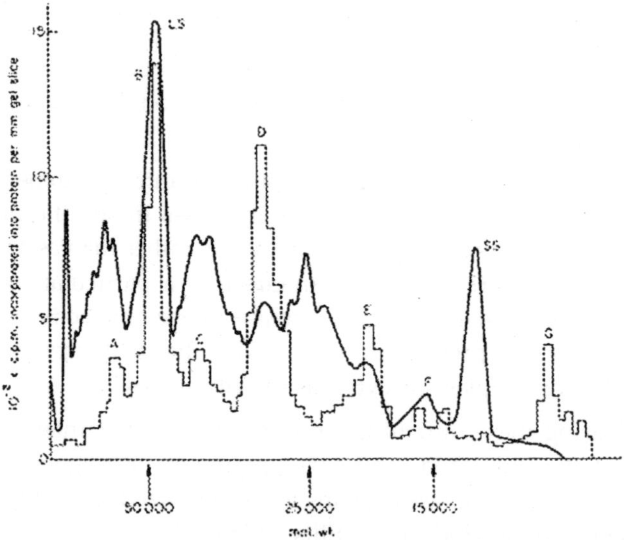

***Figure: Products of protein synthesis by chloroplasts isolated from* Pisum sativum.**

That this is true for Fraction I protein has been shown by the second approach to the problem of chloroplast ribosomal function, namely, the synthesis of specific proteins by isolated sub-cellular systems.

This is the most direct approach to this problem but it has been successful only recently because of the difficulty of isolating sub-cellular systems which will carry out complete protein synthesis with fidelity; algal systems are especially difficult in this respect. Isolated intact chloroplasts from *Pisum sativum*, however, synthesize at least six discrete proteins; similar results have been obtained with chloroplasts from *Hordeum vulgare* and *Spinacia oleracea*.

Only one of these proteins is soluble; it has been identified as the large subunit of Fraction I protein by tryptic peptide analysis (Blair & Ellis, 1973).

Recent work has shown this large subunit to be made by the free ribosomes of the chloroplast, but not by the membrane-bound ribosomes. The high proportion of chloroplast ribosomes found in leaves may thus be required because Fraction I protein is one of the most abundant proteins (Ellis, 1973).

The small subunit of Fraction I protein is not synthesized by isolated chloroplasts, but it has been detected as the product of protein synthesis by isolated cytoplasmic ribosomes from *Phaseolus vulgaris* leaves (Kekwick & Gray, 1973).

The other proteins made by isolated chloroplasts are all membrane-bound, and have so far resisted identification; it is clear they are minor components of the thylakoids. Two proteins associated with the chloroplast envelope are also labelled in isolated pea chloroplasts (Joy & Ellis, 1975). The same pattern of proteins synthesized by intact pea chloroplasts has been found in studies of protein synthesis by isolated pea etioplasts (Siddell & Ellis, 1975).

This *in vitro* approach to the study of chloroplast ribosomal function gives direct and unambiguous results, and it should be extended to species lower on the evolutionary scale to see how far the pattern found in angiosperms extends. In view of the uniformity of the size of the chloroplast genome in algae and higher plants, it seems likely that the pattern seen in isolated pea chloroplasts will be universal. However, this assumes that proteins synthesized by chloroplast ribosomes are also encoded by chloroplast DNA; this may not be true if messenger-RNA can cross the chloroplast envelope, but so far this is only a theoretical possibility.

Another area which may repay attention is the study of protein synthesis by plastids other than chloroplasts, especially proplastids and amyloplasts. It seems probable that these plastids synthesize a different range of proteins from chloroplasts. If these proteins are encoded in plastid DNA, they would account for some of the functions of the 90% of this genome which have not been identified.

Co-Operation between Chloroplast and Nuclear Genomes

It is clear from the available evidence that the chloroplast genome requires for its expression the co-operation of the nuclear genome. The best information about the details of this co-operation concerns the synthesis of Fraction I protein.

By combining the results of Wildman's genetic studies with those of the *in vitro* studies of protein synthesis, a model for the synthesis of Fraction I protein can be constructed.

In this model, the large subunit is both encoded and synthesized within the chloroplast, while the small subunit is both encoded and synthesized outside the chloroplast. This model is tidy, and requires protein, but not nucleic acid, to cross the chloroplast envelope.

This transport of protein must be on a large scale because it involves not only the small subunit of Fraction I protein but all the other proteins which, from the inhibitor evidence, are made on cytoplasmic ribosomes.

The mechanism of this transport is unknown, but it must be able to distinguish between different proteins. One suggestion is that a protein exists in the outer envelope of the chloroplast which recognizes a site common to all those proteins made on cytoplasmic ribosomes, but destined to function in the chloroplast.

Besides structural and enzymic proteins, regulatory proteins probably also cross the chloroplast envelope.

How light regulates chloroplast development is not known, but it would be reasonable to speculate in terms of proteins entering the organelle to trigger nucleic acid and protein synthesis.

Figure: Model for the co-operation of nuclear and chloroplast genomes in the synthesis of Fraction I protein. cDNA and nDNA stand for chloroplast and nuclear DNA respectively. The dashed lines indicate possible sites at which small subunits may control the synthesis of large subunits. (Reprinted from Ellis, 1975, by courtesy of Pergamon Press.)

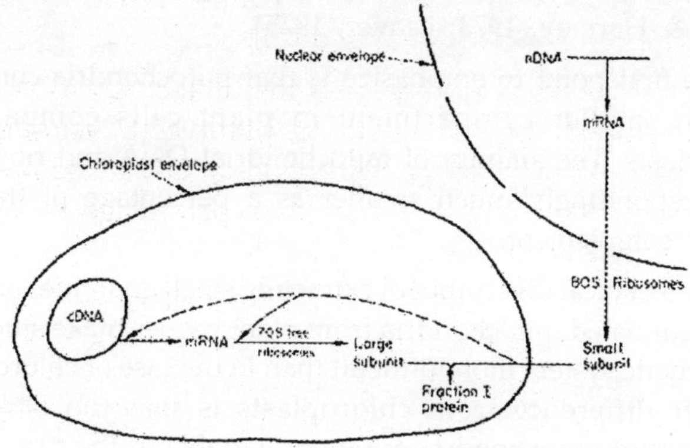

There is the possibility of proteins also passing out of the organelle, since recent work in *Chlamydomonas* has implicated chloroplast protein synthesis in the regulation of nuclear DNA replication (Blamire *et al.*, 1974).

This question of protein transport into and out of the chloroplast is a crucial one to study if we are to understand how the chloroplast and nuclear genomes interact; almost no research has been carried out in this area.

The movement of nucleic acids across the envelope remains a possibility, but no compelling evidence to support it has been published.

Another problem needing study is the nature of the mechanism which integrates the synthesis of the two subunits of Fraction I protein in different cellular compartments. The dashed lines in illustrate one possibility; the small subunit is postulated to be required as a positive factor for the initiation of either transcription or translation of the large subunit messenger (Ellis, 1975). The recent detection of the large subunit messenger should allow this possibility to be tested (Hartley *et al.*, 1975).

MITOCHONDRIAL AUTONOMY

Research on the biogenesis of plant mitochondria has lagged far behind that on mitochondria from animal and fungal cells. A summary of the position is given here in the hope of stimulating readers to undertake work in this field; recent reviews have appeared on the biogenesis of both animal and fungal mitochondria (Borst, 1971, 1972), and of those in plants (Boulter *et al.*, 1972; Leaver & Harmey, 1973; Leaver, 1975).

The first point to emphasize is that mitochondria constitute a much smaller compartment in plant cells compared to chloroplasts. The amount of mitochondrial DNA and ribosomes is correspondingly much smaller as a percentage of the total cellular complement.

The technical difficulties of extracting small quantities of intact, uncontaminated mitochondria from plant tissues make studies on their genetic system more difficult than in the case of chloroplasts. Another difference from chloroplasts is that the DNA and ribosomes of mitochondria are more variable in their properties between species, so the extrapolation of results must be done with caution.

In a range of higher plants the density of mitochondrial DNA is constant at 1.706–1.707 g cm^{-3}, whereas the nuclear DNA from the same species varies in density between 1.691–1.702 g $cm–^{-3}$. In many organisms mitochondrial DNA has been isolated in circular form; in contrast to the situation for chloroplasts, the contour length of the circles varies with the species.

Thus, animals have mitochondrial DNA circles of about 5 μm contour length; this corresponds to a molecular weight of about 10^7, or 15,000 base pairs. In Ascomycetes the contour length of circular mitochondrial DNA is in the range 20–26 μm, while in higher plants lengths of 30 *μm* have been reported.

The small size of the mitochondrial DNA in animals means that it can code for only a few components. Hybridization studies in *Xenopus* indicate that mitochondrial ribosomal RNA and at least some mitochondrial transfer RNA are encoded in mitochondrial DNA.

The remaining mitochondrial DNA is sufficient to code for only about ten proteins each of molecular weight 50,000. However, kinetic complexity measurements on mitochondrial DNA from higher plants indicate that it has between six and ten times the coding potential of animal mitochondrial DNA.

This observation raises the possibility that mitochondria from plants possess a higher degree of autonomy than those from animals, and this reinforces the plea that more research be carried out with plant cells.

The sedimentation coefficients of mitochondrial ribosomes and their component ribosomal-RNAs vary greatly with the species.

Animal mitochondria contain so-called 'miniribosomes' which sediment between 55s and 60s, while fungal mitochondria contain 70–74s ribosomes. Leaver and Harmey (1972) have shown that mitochondrial ribosomes from several higher plants sediment at 77s–78s, while their major RNA components sediment at 25s and 18s.

Although mitochondrial ribosomes differ in size in different species, they are all similar in that protein synthesis by them is sensitive to the same set of antibiotics which inhibit prokaryote and chloroplast ribosomes.

Protein synthesis by isolated mitochondria has been studied extensively in animals and fungi, but hardly at all in higher plants. Several proteins of the inner, but not the outer, mitochondrial membrane are synthesized by animal and fungal mitochondria.

These proteins have been identified as subunits of the ATPase complex, cytochrome *b*, and cytochrome *c* oxidase; the other subunits of these insoluble complexes are synthesized by cytoplasmic ribosomes.

This evidence suggests the synthesis of mitochondrial membranes requires the co-operative activity of both mitochondrial and nuclear genomes. This idea of the co-operation between the several genetic systems in eukaryotic cells is already familiar to the reader. The same problem of the transport of proteins across the bounding membranes applies to mitochondria as well as to chloroplasts. In contrast to chloroplasts, mitochondria from animal and fungal cells synthesize no soluble proteins; whether higher plant mitochondria synthesize any soluble proteins is unknown.

No protein has so far been rigorously identified as being encoded in mitochondrial DNA in any organism. A simple presumption is that those proteins synthesized by isolated mitochondria are encoded in mitochondrial DNA. It is likely that the integrative controls between nuclear and mitochondrial genomes in higher plants differ from those in animals which contain less than one sixth as much DNA. Why do plant mitochondria apparently contain so much more genetic information to perform the same functions as animal mitochondria? Answering this question is a challenge which must be taken up if our understanding of the origin of this organelle in plants is to improve.

3

ENZYME LEVELS AND ACTIVITY REGULATION

INTRODUCTION

The first theory of metabolic control was introduced into plant physiology in 1905 by F. F. Blackman and has become known as the 'Law of Limiting Factors' which states that 'when a process is conditioned as to its rapidity by a number of separate factors the rate of the process is limited by the pace of the slowest reaction'.

It is not possible to derive a rate equation from this statement and discussions of the 'Law' tend to be by analogy, e.g. the strength of a chain is the strength of its weakest link. Experimentally, the statement has led to the fruitless search for metabolic master reactions. Mathematically, we can derive rate equations for multistep reactions and establish that *every* step contributes to the overall rate. Theoretically Blackman's statement is a denial of the steady state; if a metabolic sequence is in a steady state then the concentration of each intermediate is constant and the individual reactions proceed at the same rate—hence no reaction can be described as the slowest.

Pacemakers

If in a metabolic sequence all reactions are proceeding at the same pace then a single reaction may be almost entirely responsible for determining that pace; such a reaction has been called a 'bottleneck' or pacemaker and much work in metabolic control is directed towards the identification of such control points. Krebs and Kornberg (1957) in a consideration of pacemakers stated 'there is a principle which may guide the search for pacemakers. As pacemakers are reactions of variable rate, the level of substrate concentration of the pacemaker must vary inversely with the rate: it must increase when the reaction rate decreases'. It should be possible to take published data for a metabolic pathway, apply the Krebs-Kornberg principle and so determine which reaction is the pacemaker. It turns out that in some cases the Krebs-Kornberg principle cannot be applied. For example in some tissues when the rate of glycolysis is increased not a single compound shows the expected reduction in concentration.

Occam's Razor

William of Occam, the mediaeval controversialist, formulated a procedural rule, that when a number of possible solutions can be proposed for a problem, one should accept the simplest solution until it is shown to be untenable. The Krebs-Kornberg principle is a simple solution to a complex problem and in certain cases it yields a valid solution. However, when it is not applicable we must examine more complex solutions.

Systems Properties

The rate equation for a metabolic pathway includes parameters for all the enzymes and variables for all the metabolites (Waley, 1964). The flux through the pathway is a systemic property in which all the parameters interact as a system. If a single parameter is altered, say the activity of a single enzyme, then the whole system responds and adjusts to that change. The extent to which an enzyme can be considered a control step is the extent to which a fractional change in its activity produces a fractional change in the flux through the system. The ratio has been termed the sensitivity coefficient for that step (Kacser & Burns, 1973). If a 1 % change in activity of an enzyme produces a 1% change in flux through the system, then the sensitivity coefficient is 1 and the enzyme must be fully controlling the flux and is a pacemaker. However, the sum of all the sensitivity coefficients for a metabolic sequence is equal to unity and only in special circumstances can a single pacemaker be identified. That the sensitivity co-efficient of a particular enzyme is a system property, only in part determined by its own parameters, can be intuitively understood by an example. In a given situation an enzyme may have a very small sensitivity coefficient and contribute little to the overall control of the system. If however the activity of this enzyme is drastically reduced its sensitivity coefficient may be increased and it may contribute significantly to the overall flux. Since the sum of sensitivity coefficients must always equal unit the sensitivity coefficients of all the other enzymes must have changed despite the fact that their parameters are unchanged.

Much work on metabolic control is concerned with the identification of control points. Since each reaction contributes to the overall control a sense of judgement is necessary to identify major control points and it must be remembered that control may pass from one reaction to another as the flux through the system changes.

THE IDENTIFICATION OF CONTROL POINTS

Equilibrium Considerations

Bücher and Russmann (1964) have proposed a model for metabolic control based on a controlled waterway.

The flow of water through the system is controlled by the flood gates. With a flood gate up, there is a negligible gradient across the reservoir and a slight flow of water over the dam. In metabolic terms the reaction represents the reservoir and can be considered as a fast reaction close to equilibrium. The reaction B® C represents the waterfall and is a slow reaction far from equilibrium (*D* G is large).

When the flood gate is lowered a slight gradient is produced across the reservoir and there is a rapid flow of water over the dam, which could be utilized to perform work. In metabolic terms, as the flux through the system increases, the reaction moves away from equilibrium and the controlled reaction B ® C increases in rate and simultaneously moves towards equilibrium.

The model thus identifies control reactions as reactions far from equilibrium and which move towards equilibrium as the flux increases. Since *D* G for such reactions is large, control points will tend to be associated with reactions involved in the production of useful work by a coupled reaction.

Figure: Schematic representation of a regulated waterway. The solid line represents the situation when the flood gates are raised and there is little flow through the system. The dashed line represents the increased flow when the flood gates are lowered.

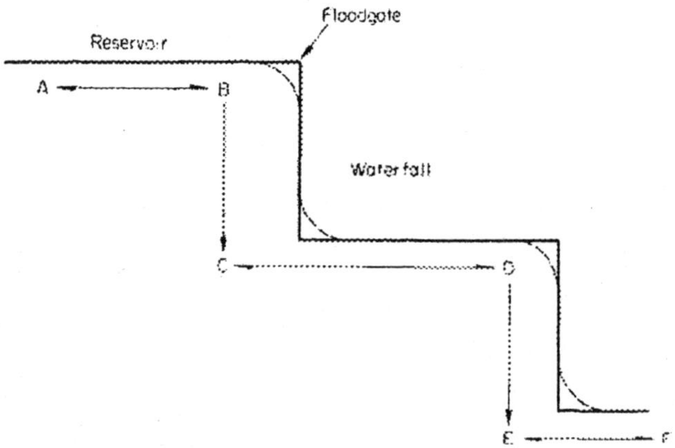

However, the useful work function is not an essential feature of the model and in the absence of a coupled reaction to utilize the large *D* G, the loss of energy can be considered as the price paid for control.

The Crossover Theorem

The crossover theorem states that: the variations of the concentrations of the metabolites before and after an enzyme which is a control point have different signs.

It will be noted that this is only a formal statement of the situation at the waterfall in the Bücher and Russmann analogy, i.e. when the flux increases the concentration of the substrate falls and the concentration of the product increases.

Figure: Crossover plot of glycolytic intermediates following transfer of aged carrot discs from air to nitrogen (Data of Faiz-ur-Rahman, Trewavas & Davies, 1974). Intermediates were measured before transfer and 6 minutes after transfer from air to nitrogen.

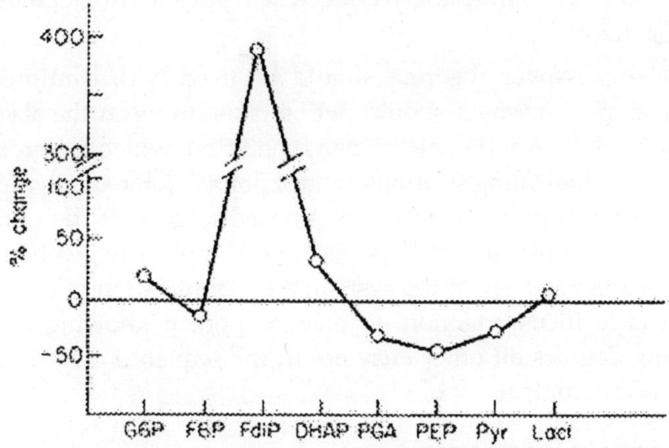

However, the theorem is widely used, possibly because it presents data in a graphical form. Metabolic intermediates are plotted on the x coordinate in the sequence in which they are formed in the metabolic pathway and the concentration of each substrate, expressed as a percentage of the initial value is plotted on the y coordinate. An example is shown in Fig. 12.2 for the situation usually referred to as the 'Pasteur Effect'. Pasteur observed that 'oxygen inhibits fermentation' so that when carrot discs are transferred from air to nitrogen glycolysis increases with associated changes in the levels of intermediates. These changes are plotted for a period some 6 minutes after transfer from air to nitrogen.

The positive crossover observed between fructose-6-phosphate and fructose diphosphate suggests that phosphofructokinase is a control point. There is some confusion concerning the interpretation of negative crossovers. Some authors argue that a negative crossover does not indicate a control point. However, it must be remembered that when the flux through a system is changed, control can pass from one point to another.

Thus in the case illustrated in Fig. 12.2 the increased flux through the glycolytic system involves an increase in activity of phosphofructokinase but with the increased rate of glycolysis, phosphoglycerate is removed more rapidly than it is formed, suggesting that a bottleneck exists between dihydroxyacetone phosphate and phosphoglycerate when phosphofructokinase is highly active.

The crossover theorem should be used with caution; the absence of a crossover should not be taken to mean the absence of a control point and there are many conditions which can produce crossovers indicating spurious control points. Before applying the crossover theorem, the reader is advised to consult the critique offered by Heinrich and Rapoport (1974) and to remember that control is a property of the system; the identification of a control point is to focus attention on one component knowing that to varying degrees all other enzymes in the sequence contribute to the overall control.

CONTROL MECHANISMS

Having identified possible control points it becomes necessary to determine the mechanism of control. Over a relatively long time scale control may be achieved by changing the amount of an enzyme but over a short time scale control is likely to be achieved by changing the activity of the enzyme.

Control by Product Inhibition

Every enzyme has a certain affinity for its products and if the enzyme has a high affinity for its product then product inhibition will be pronounced. It appears likely that this form of control is restricted to some synthetic routes and to minor metabolic sequences.

Control by Negative Feedback

The discovery of negative feedback in metabolic systems appears to have been made by Dische in 1940 who noted that 3-phosphoglycerate inhibits the phosphorylation of glucose in red cells and he proposed that this inhibition might play a regulatory function in glucose metabolism. His paper did not receive the attention it deserved and the idea of feedback had to be rediscovered.

In 1955 Roberts and his coworkers published the result of their extensive studies of the metabolism of *E. coli* grown on a medium containing ^{14} C-glucose. They found that the addition of one of a number of amino acids (known to be products of specific synthetic pathways) to the growth medium resulted in the virtually complete inhibition of incorporation of ^{14} C into that amino acid.

In 1956 Umbarger reported that isoleucine inhibited threonine deaminase—an enzyme which initiates the set of reactions leading to isoleucine biosynthesis. In the same year Yates and Pardee observed that cytidine monophosphate inhibits aspartate transcarbamylase—the first reaction of the pathway leading to the biosynthesis of the pyrimidines. These results offered a biochemical explanation for the end-product inhibition observed by Roberts with intact bacteria.

In 1959 Sir Hans Krebs introduced the first symposium to be held on metabolic control. Ideas on the regulation of carbohydrate and amino acid metabolism were discussed and the general ideas of feedback formulated. Control by negative feedback is an engineering concept and familiar examples include governers of steam engines, thermostats and automatic volume control in amplifiers.

The limitations of the analogy between electrical feedback amplifiers and metabolic reactions have been stressed by Britton-Chance. However, as with other terminology developed for use with metabolic control systems (see discussion of allosteric enzymes) the tendency has been to dispense with rigorous definitions and the general definition of negative feedback is 'increased output decreases the input'.

Patterns of Control

The most frequently observed pattern of control is when the end. product of a metabolic sequence inhibits the first reaction belonging to that sequence.

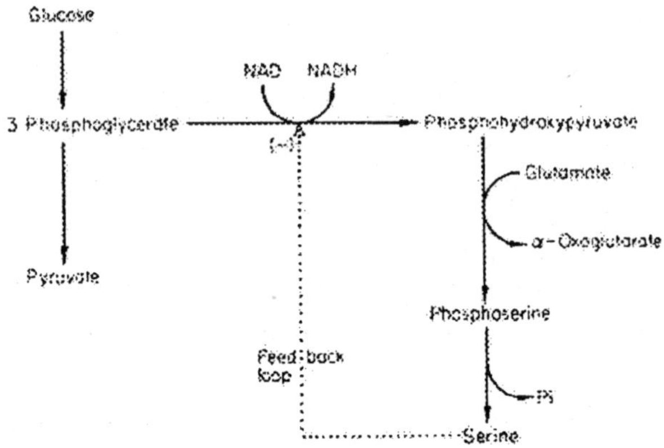

Figure: Reactions involved in the biosynthesis of serine.

Serine inhibits 3-phosphoglycerate dehydrogenase and so controls its own biosynthesis.

This simple mechanism is effective, presumably because a relatively small amount of the flux through the glycolytic pathway is directed towards serine and thus variations in the rate of serine production will not seriously perturb the glycolytic flux.

Control of Branched Pathways

Inhibition of the first step in a metabolic pathway by an end-product leads to special problems in the case of branched biosynthetic pathways. Consider the sequence :

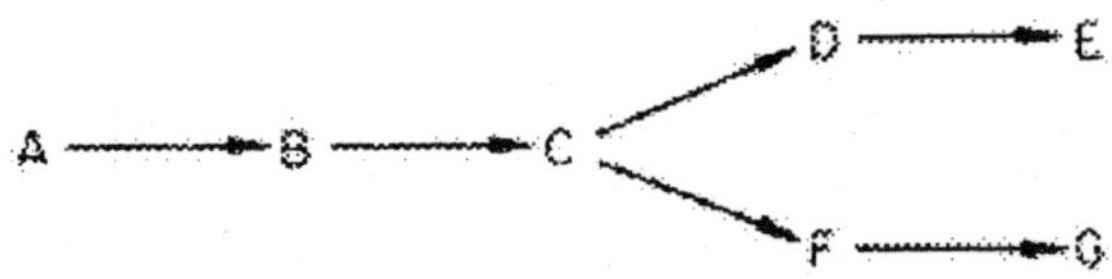

It the first common step (A® B) is inhibited by either or both end-products, then an excess of one could inhibit the step A® B and lead to a deficiency of the other end-product. Nature has evolved several mechanisms which avoid, to varying degrees, these difficulties.

Stadtman (1970) has classified these mechanisms; here we discuss a few examples which have been studied in plants.

The Aspartate Family

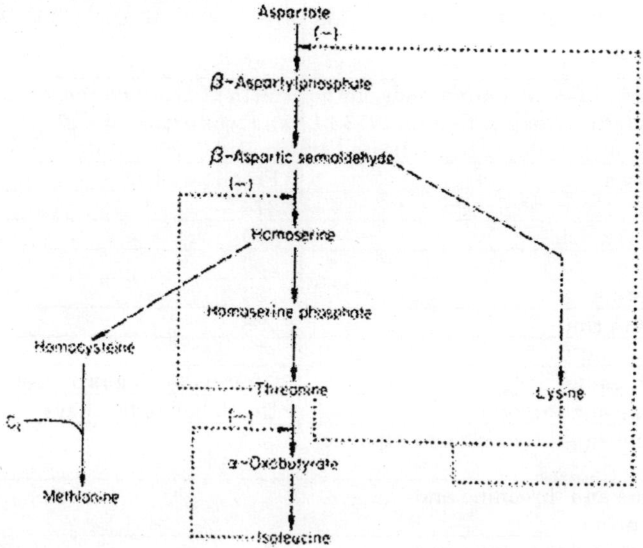

Figure: Reactions involved in the biosynthesis of the aspartate family of amino acids. ® Single step reaction; —® multi-step reaction; —® feed back loop.

Aspartokinase which catalyses the first reaction of the pathway is inhibited by lysine and threonine in a concerted or synergistic manner. The term concerted feedback inhibition has been used by Stadtman to describe cases of inhibition by two end products when the single end products produce no inhibition. The term synergistic inhibition is used when the total inhibition is much greater than the sum of their independent effects. The aspartokinase from plants appears to be inhibited by low concentrations of lysine but not by low concentrations of threonine—when both are present a much greater inhibition is observed.

Although this effect would seem best to fit Stadtman's definition of synergistic inhibition it has been called concerted inhibition. Whatever the terminology, this mechanism does not involve methionine and is thus only a partial solution to the control of the branched pathway. The non-involvement of methionine means that in conditions where threonine and lysine.

These imperfections in the control mechanism can be invoked to explain the effect of amino acids on the growth of *Lemna minor* (Table).

Table . Effect of amino acids on the growth of *Lemna minor* in 10 days (after Wong & Dennis 1973.) *Lemna* was grown in a growth medium containing the indicated amino acid (0.25 mM).

Addition	Frond number
None	373
Aspartic acid	400
Lysine	230
Threonine	84
Homoserine	98
Isoleucine	73
Methionine	After 5 days signs of death
Lysine and threonine	Death before five days
Lysine and threonine and aspartic acid	,,
Lysine and threonine and isoleucine	,,
Lysine and threonine and homoserine	203
Lysine and threonine and methionine	48

The relatively small inhibition observed with lysine can be interpreted in terms of inhibition of aspartokinase (but see section on *metabolic interlock*). The inhibition produced by threonine is consistent with the inhibition of homoserine dehydrogenase which would prevent the flow of carbon to methionine. The inhibition produced by homoserine is difficult to explain since as far as known it does not inhibit any of the enzymes involved. Furthermore in some plants e.g. peas, homoserine is produced in large quantities and transported in the phoem.

The inhibition produced by isoleucine is consistent with a control pattern known as *sequential inhibition.* Isoleucine has been shown to inhibit threonine deaminase which would be expected to lead to an accumulation of threonine which in turn inhibits homoserine dehydrogenase and so reduces the flow of carbon to methionine. The inhibition produced by methionine is difficult to explain in terms of the reactions.

The pronounced inhibition observed with lysine and threonine is consistent with their synergistic effects on aspartokinase producing a reduction in the flow of carbon to methionine, which is reinforced by the inhibition of homoserine dehydrogenase by threonine. Aspartic acid and isoleucine are unable to reduce the inhibition but methionine and homoserine are able to do so, presumably by supplying methionine directly or indirectly.

Aromatic Biosynthesis

The control of aromatic amino acid biosynthesis has been extensively studied in fungi and bacteria and it is clear that nature has evolved many solutions to the problem of control in branched pathways.

Enzyme Multiplicity

In this control mechanism the first common step is catalysed by two or more enzymes which are under feedback control by compounds formed after the branching point. The control of chorismate mutase being particularly complicated. Three isozymes have been demonstrated in plants (Woodin & Nishioka, 1973); CM_1 and CM_3 are inhibited by phenylalanine and *activated* by tryptophane. CM_1 and CM_2 are inhibited by caffeic acid and chlorogenic acid whilst CM_3 is inhibited by ferulic acid.

Enzyme Aggregation

A number of fungi possess an aggregate of the five enzymes necessary for the conversion of 3-deoxy-D -heptulosonate-7-phosphate to enolpyruvyl-shikimate-5-phosphate. In *Neurospora* the five enzymes form a complex (M.W. 230,000) coded by the *arom* gene cluster.

This large complex has not been demonstrated in higher plants, but a bifunctional enzyme consisting of dehydroquinase and shikimate dehydrogenase has been isolated (Boudet, 1971). *Neurospora* contains two dehydroquinases, the one in the aggregate is involved in an anabolic sequence, the other maps outside the *arom* cluster and is thought to function as a component of a catabolic sequence.

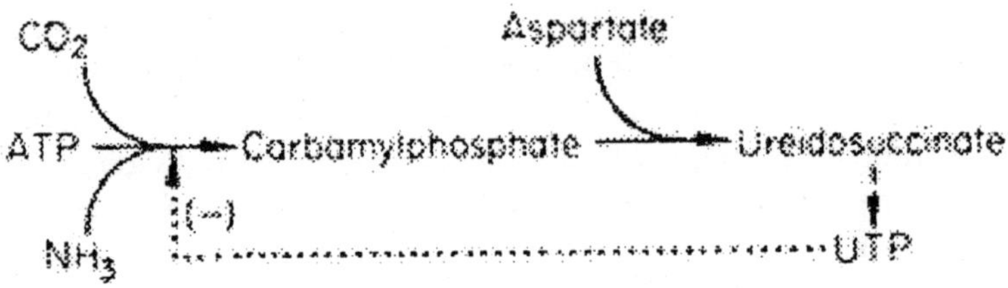

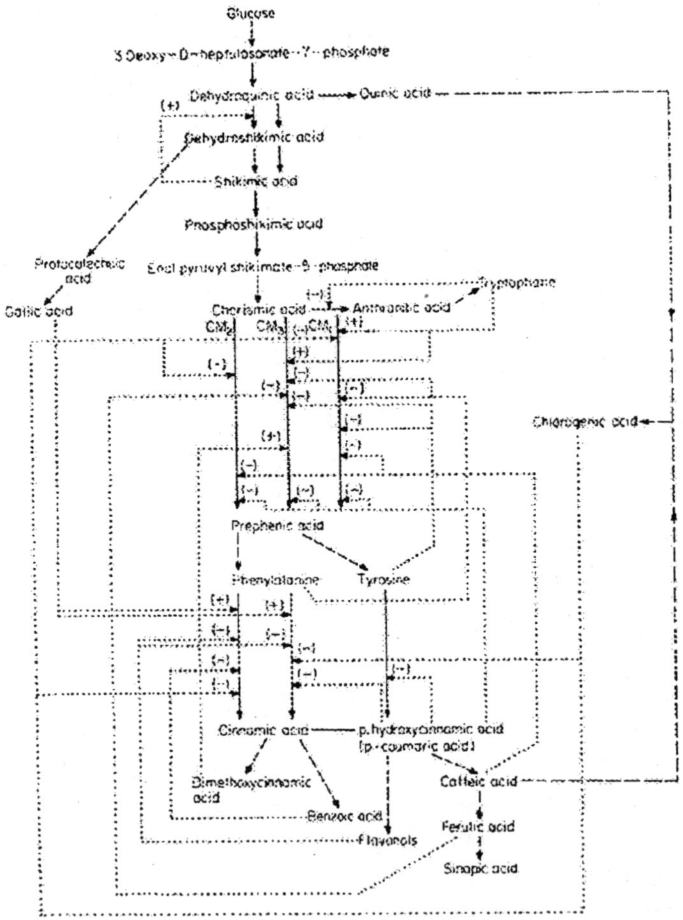

Figure: Reactions involved in the biosynthesis of aromatic compounds. ® Single step reactions; —® multi-step reactions; —® feed back loop.

The physiological significance of the multienzyme cluster could be to separate the degradative and synthetic routes. This could be achieved if the dehydroquinic acid which is formed in the aggregate is not released as a free product. This idea has become known as *metabolic channelling* and in a number of cases it has been shown that metabolites involved in channelling do not leave the enzyme surface. For example fungi possess a bifunctional enzyme consisting of carbamyl phosphate synthetase and aspartate transcarbamylase.

When the bifunctional enzyme is synthesizing ureidosuccinate from $^{14}CO_2$, the addition of carbamyl phosphate does not dilute the incorporation of ^{14}C into ureidosuccinate, showing that carbamyl phosphate formed by the double enzyme does not equilibrate with free carbamyl phosphate.

Higher plants contain two dehydroquinases one forming a bifunctional association with shikimate dehydrogenase, the other being activated by shikimic acid. Such a system involving a specific regulation of isoenzymes and molecular compartmentalization may function to control the partitioning of dehydroquinic acid between the pathways to phenolcarboxylic acids and aromatic amino acids.

Metabolic Interlock

The reactions and feedback loops shown in Figure above give some idea of the complexity of control within a metabolic pathway. When the feedback loops of other metabolic pathways are taken into consideration it seems highly likely that the control must be interlocked. As an example, consider the reactions involved in the biosynthesis of methionine.

The active C_1 group is generated in the serine hydroxymethyl transferase reaction as methylene tetrahydrofolate. After reduction to methyl THFA the methyl group is transferred to homocysteine to form methionine. Homocysteine is derived from the aspartate family and so is lysine. The metabolism of the aspartate family and the serine glycine family are thus interlocked *via* lysine, which inhibits both aspartokinase and serine hydroxymethyl transferase.

Figure: Reactions involved in the biosynthesis of methionine. ® Single step reaction; —® multi-step reaction;...® feed back loop.

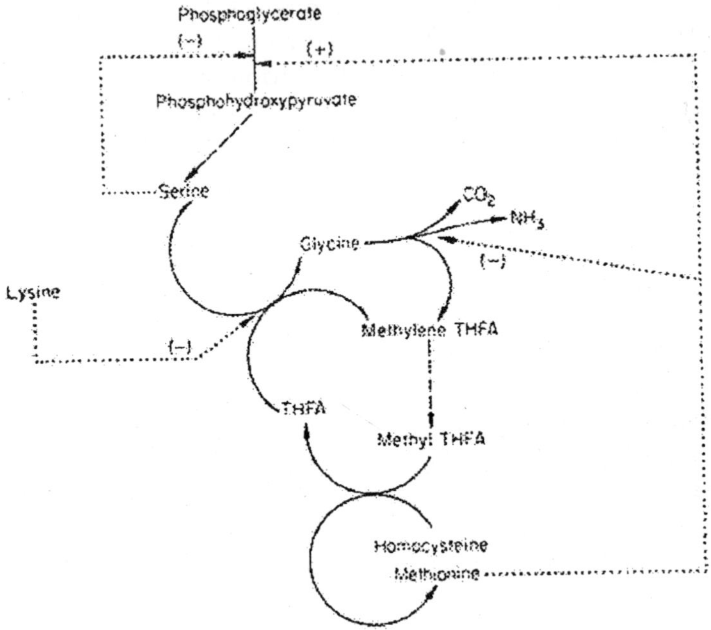

Enzymes as Control Elements

Until the discovery of feedback inhibition in 1956 enzymes were considered as biological catalysts and most enzymologists were concerned with the properties of the active site. When it became apparent that enzyme activity was under the control of specific metabolites, which were structurally unlike the active site, it became necessary to postulate a control site to which metabolites could bind and modulate the activity of the enzyme. Recognizing the lack of structural similarity between the substrate and the effector the term allosteric site was introduced (Monod, Changeux & Jacob, 1963). Subsequently Monod, Wyman and Changeux (1965) noted that many regulatory enzymes with specific allosteric sites exhibit cooperative-type kinetics and they developed a concerted transition hypothesis to explain the observed kinetics. Many workers assume that any enzyme showing sigmoidal kinetics is an allosteric enzyme and some workers imply that the mechanism of activation of all allosteric enzymes follows the mechanism postulated by Monod *et al.*, (1965). The purist may object but the original precise definition has been replaced by a broader but less precise usage.

Allosteric Enzymes

Before 1956, the kinetics of numerous enzymes were shown to follow the rate law of Michaelis and Menten giving a hyperbolic plot of **v** against **S**. An examination of enzymes which were known to be subject to feedback regulation showed that in some cases the relationship between **v** and **S** was sigmoid. In other cases the addition of a feedback inhibitor led to a change from a hyperbolic to a sigmoid relationship.

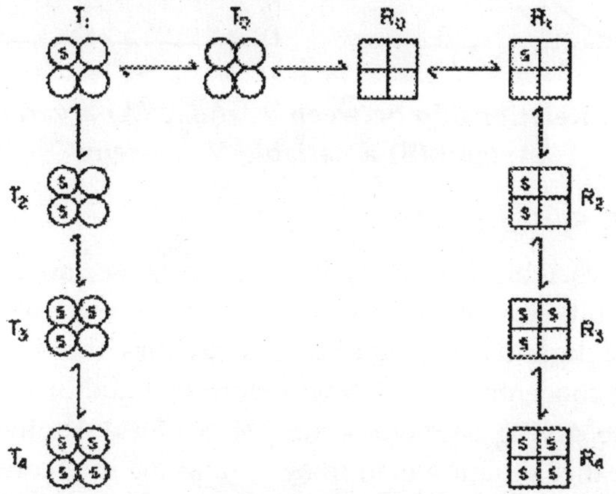

Figure: Illustration of the Monod model of an allosteric enzyme.

Monod *et al.*, (1965) proposed a simple model to account for the observed kinetics in which they took account of the observation that allosteric enzymes are composed of subunits. They assumed that the protein subunits were arranged in such a way that they occupy equivalent positions so that the enzyme must have at least one axis of symmetry. They further assumed two interconvertible states—R and T—each maintaining the symmetry principle. The reader is referred to the original article by Monod *et al.*, (1965) for the derivation of the rate law. However a consideration of the illustration leads to a basic understanding of the underlying causes of sigmoid kinetics. Two types of allosteric behaviour may be distinguished—one the variable *K* system in which the two states of the enzyme have different affinities for the substrate, the other, the variable *V* system in which the *R* state is catalytically more active.

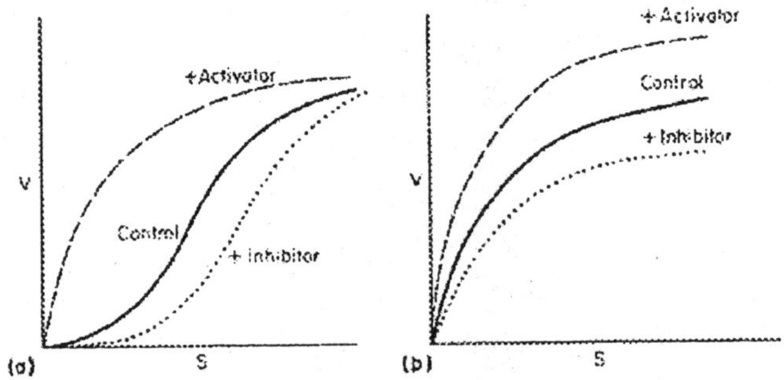

Figure: Relationship between v and** s. **(A) a variable** K **system; (B) a variable** V **system.

Variable K Systems

In the variable *K* system, the *R* form is assumed to have a higher affinity for **S** than the *T* form of the enzyme. If in the absence of ligands, the equilibrium favours the *T* form then increasing concentrations of S will shift the equilibrium towards the *R* state giving sigmoid **v** against **S** plots. Another way of looking at this situation is to imagine that the interconversion of the *R* and *T* forms can be blocked and the enzyme isolated as the *R* or *T* form. The *T* form would give a normal Michaelis-Menten curve with a high $Km^{(T)}$ and the *R* form would also give a Michaelis-Menten curve but with a low $Km^{(R)}$. Allowing *R* and *T* to be interconvertible allows *Km* to vary.

If an activator binds preferentially to the *R* state it will tend to put the enzyme entirely in the *R* state and so remove the sigmoidicity of the **v** against **S** plot. Similarly if an inhibitor binds preferentially to the *T* form it will increase the sigmoidicity.

Variable V Systems

In the variable *V* system the *R* form is assumed to be catalytically more active than the *T* form though both forms have the same affinity for the substrate so that the **v/S** plot is hyperbolic in the presence or absence of activators and inhibitors. An activator is assumed to have a greater affinity for the *R* form than for the *T* form whilst the reverse holds for an inhibitor.

Alternative Models for Allosteric Enzymes

This superficial consideration of the Monod model concentrates on only one aspect of the model–the two configurational states.

To explain higher order reactions the Monod model includes some highly restrictive assumptions which produce mathematical simplicity but are nevertheless based on the fundamental properties of protein structure.

An alternative model proposed by Koshland (1970) assumes an induced fit so that the protein subunit undergoes a conformational change when it binds a ligand. A simple sequential model may be illustrated:

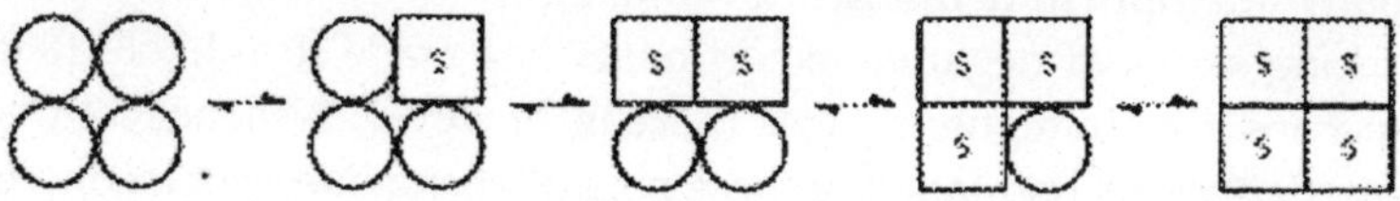

This simple model is not a necessary part of the induced fit model (which in fact suggests a diversity of forms of association between enzyme and ligand—25 for a tetrameric protein). Nevertheless it simplifies the mathematical treatment and in many cases provides a close fit to experimental values.

Kinetic Constants for Allosteric Enzymes

When an electronic engineer constructs a control circuit he needs to know the characteristics of the components—the amplification factor, grid bias, etc. Similarly if we wish to understand the control of a system we need to know the characteristics of its components, that is, the kinetic constants of the enzymes. Clearly a variable *K* system cannot be defined by a *Km* value nevertheless the same operation value can be used, that is the concentration of substrate producing half maximum velocity which is in this case designated $\mathbf{S}_{0.5}$. More generally the term ($X_{0.5}$) is used to designate the ligand concentration at half saturation.

The constant *N* is operationally the order of the reaction, its physical meaning may relate to the number of substrate binding sites on the enzyme or to the degree of cooperativity of binding. Whatever the meaning of *N* it can be estimated by means of a Hill plot (Fig. 12.9). Another kinetic term, R_s is the ratio between the 90% and 10% saturation values. R_s will be 81 for all curves following Michaelis-Menten kinetics and less than 81 for cases of positive cooperativity–a few cases of negative cooperativity are known and these give R_s values greater than 81.

CONTROL MECHANISMS INVOLVING CHANGES IN THE AMOUNT OF ENZYME

The brilliant work of Jacob and Monod (1961) on the regulatory mechanisms governing *b* -galactosidase activity in *E. coli* led to many attempts to demonstrate similar control systems in plants. A long series of negative experiments has made it unlikely that operons exist in higher plants. There is, however, evidence for the interdependent regulation of some groups of enzymes. For example in parsley and soybean cell cultures, the three enzymes catalysing the synthesis of *p* -coumaroyl CoA from phenylalanine show strictly concomitant changes in activity. Large changes in activity are initiated by light but there is no satisfactory explanation for the strict coordination in activity. Mohr (1972) has suggested that genes can be divided into four classes—active genes which function in the same way in light and dark, inactive genes which function only at special times in the life history of the plant (e.g. flowering genes), potentially active genes whose activation is under the control of the phytochrome system and repressible genes which can be repressed by the phytochrome system. Positive photoresponses are identified with gene activation and negative photoresponses with gene repression. This model is so basic that it must contain some truth, but it offers no explanation of the mechanism of regulation other than to identify regulation as occurring at the gene level. Much of the evidence for Mohr's model is concerned with the increase in phenylalanine ammonia lyase which occurs when mustard cotyledons are exposed to far-red light. Recently however, it has been suggested (Attridge *et al.*, 1974) that this increase in activity is due to an activation of pre-existing enzyme rather than synthesis.

Following the success in understanding the regulatory system in prokaryotes it was clearly sensible to see if the system was applicable to plants. It now appears that the regulatory systems are very different, but the technical vocabulary used in the study of prokaryotes is still used in studies with plants. It should be noted that terms such as repression and derepression have precise meanings in relation to molecular models of regulation in prokaryotes, but when applied to plants, these terms should not be taken to imply a mechanism similar to that found in bacteria.

Table . End-product control of enzyme activity.

End-product	Tissue	Enzyme showing reduction in activity
Inorganic phosphate	Wheat scutellum	Phytase
Glucose	Sugar cane	Invertase
Glucose	Wheat embryo	Glucose-6-phosphatase and fructose-1, 6-diphosphatase
Arginine	Soybean culture	Arginosuccinate synthetase
Ammonia	*Lemna* plants	Nitrate reductase

Many cases are known where a metabolite reduces the level of an enzyme involved in its metabolism. Most of these cases are reported as examples of repression but the mechanism involved varies from case to case—for example arginine probably exerts its effect by stimulating proteolysis, the effect of ammonia may be due to proteolysis, though the formation of a specific protein inactivator has been proposed, and the effect of glucose on invertase has been reported as due to the destruction of m-RNA required for invertase synthesis.

The regulation of enzyme levels may involve one or more of the following:

(a) Degradation

(b) Inactivation

(c) Activation

(d) Synthesis

When examining a particular system it is necessary to establish which control system is involved and to measure its rate—this is frequently surprisingly difficult.

Protein Turnover and Degradation

The view that plant proteins may undergo a cycle of synthesis and degradation was first proposed by Gregory and Sen in 1937. The terminology is somewhat confused; protein turnover implies synthesis *and* degradation but usually only degradation is measured.

Three methods are available for determining the overall rate of protein degradation.

(a) Incorporation of ^{14} C-amino acids

The tissue is pulse-labelled with a ^{14} C-amino acid and the incorporation into protein measured. During the chase period the rate of protein degradation is estimated from the fall of radioactivity in the protein. The method has two serious disadvantages:

(1) The recycling of protein amino acids—that is the reincorporation of a ^{14} C-amino acid released by degradation tends to underestimate the rate of degradation. This disadvantage can be minimized by using a ^{14} C-amino acid which is rapidly metabolized;

(2) The existence of pools of amino acids means that the specific activity of the amino acid at the site of protein synthesis is unknown. Furthermore the incorporation of a labelled amino acid into a metabolic pool during the pulse period may be followed by a slow release of ^{14} C-amino acid during the chase. This disadvantage can be

minimized by using an amino acid whose pool size is small, e.g. leucine.

(b) Measurement of the specific activity of aminoacyl t-RNA

Trewavas (1969) has developed a method in which the specific activity of the amino acid entering protein during a labelling experiment is determined by isolating the appropriate aminoacyl t-RNA.

(c) Measurement of the incorporation of ^3H

Humphrey and Davies (1975) established that when *Lemna* fronds are exposed to ^3H_2O, the tritium rapidly enters and equilibrates with H on the *a* -carbon atoms of amino acids. Protein synthesized during the exposure to ^3H_2O is thus labelled in a stable position. When the plants are transferred to H_2O the amino acids rapidly lose their ^3H, the ^3H in protein is stable, but as soon as the protein is degraded to amino acids the ^3H exchanges with H_2O. Thus the rate of protein degradation can be determined simply by measuring the loss of ^3H from protein.

The method depends upon the speed with which transaminases catalyse the exchange reaction between water and the hydrogen on the *a* -carbon atom of amino acids. If it can be established for a particular tissue that this reaction is fast relative to the rate of protein synthesis the method overcomes problems of pools and recycling. The method can in principle be applied to a specific enzyme, but this requires the isolation of the pure enzyme.

DEGRADATION OF SPECIFIC ENZYMES

If the degradation of proteins were a random process then a low rate of protein hydrolysis would almost completely destroy overall enzyme activity. This is because a 5% per day rate of degradation means that *each* enzyme molecule would be subject to an average degradation of 5% and for many enzymes this would mean a complete loss of catalytic activity. Consequently we must look for specificity in degradation. A NADH-nitrate reductase-inactivating enzyme has been isolated from maize roots which appears to have a high degree of specificity (Wallace, 1974). The enzyme has been purified 460-fold, the purified preparation has proteolytic activity but no peptidase activity and the capacity to inactivate nitrate reductase is blocked by phenylmethylsulphonyl fluoride suggesting the involvement of a serine residue at the active site.

The Measurement of Enzyme Half-Lives

If protein synthesis is blocked by an appropriate inhibitor and at various time intervals the level of enzyme is measured, we can determine the fall in activity of the enzyme. If we assume that the inhibitor completely blocks protein synthesis without interfering with the reactions involved in inactivating or degrading the enzyme we can estimate its half-life.

Table . Half-lives of some plant enzymes.

Enzyme	Tissue	Half-life
Vacuolar invertase	Sugar cane	2 hrs.
RNase	Sugar cane	2–4 hrs.
Nitrate reductase	Maize seedlings	4.2 hrs.
Phenylalanine ammonia-lyase	Mustard seedlings (dark)	6 hrs.
Cellulase	Pea epicotyls	24–30 hrs.
Isocitrate lyase	Water melon seedlings	2–3 days
Malate synthetase	Water melon seedlings	2–3 days
Peroxidase	Sugar cane	Stable

This method is open to criticism since in a number of cases inhibitors of protein synthesis have been shown to interfere with enzyme degradation. Thus cycloheximide stops the decline in phenylalanine ammonia lyase which follows the light induced elevation of the enzyme in potato discs and also the decline in phosphoglucomutase which occurs on slicing potato tubers. In these cases, cycloheximide is assumed to block the synthesis of the specific proteolytic enzymes necessary for the removal of the lyase and the mutase.

INACTIVATION BY PROTEIN-PROTEIN INTERACTION

The modulation of enzyme activity by the interaction of protein subunits is widely recognized. A number of reports suggest that interaction between different proteins may result in enzyme inactivation. For example the naturally occurring inhibitor of potato tuber invertase appears to be a protein which forms an essentially undissociable complex with invertase. The inactivation of *Lemna* nitrate reductase has been attributed to protein-protein interaction and ammonium adapted plants are said to contain a protein which inhibits nitrate reductase.

However, the most clearly established case is the interaction between ornithine transcarbamylase and arginase in *Saccharomyces* (Wiame *et al.*, 1973). The activity of ornithine transcarbamylase was shown to decline following the addition of arginase. It was established that arginase binds to the ornithine transcarbamylase and inhibits its activity. The binding is on a one to one basis and does not reduce the arginase activity. This interaction has not been observed in other organisms.

Chemical Modification

A number of enzymes are inactivated by low concentrations of pyridoxal phosphate at physiological pH. The mechanism of inactivation is the formation of a Schiff base between pyridoxal phosphate and an amino group of the protein, usually the *e* -amino group of lysine.

Glutamate dehydrogenase and aldolase have been shown to be inactivated by pyridoxal phosphate and to be protected by their substrates, suggesting that lysine residues are involved in the active centre of these enzymes. It is difficult to assess the physiological significance of these reactions; the concentration of free pyridoxal phosphate in plants is low but the ratio pyridoxal phosphate: pyridoxamine phosphate changes with the nitrogen status of the plant and the possibility of this ratio monitoring the nitrogen status of the plant and controlling the activities of key enzyme has been discussed.

Carbamyl phosphate also reacts with *e* -amino groups of lysine, the rate of carbamylation varying from protein to protein.

Glutamate dehydrogenase is readily inactivated by carbamyl phosphate but it is doubtful that carbamylation of enzymes has physiological significance in plants.

Enzyme catalysed chemical modification is well documented in animals and bacteria and among the important examples we note glycogen phosphorylase and glycogen synthetase of *E. coli* which is adenylated by ATP. Adenylation has not been observed in plants but phosphorylation, methylation and acetylation of proteins has been demonstrated. The extent to which these modifications are important regulatory mechanisms remains to be determined.

The classification of chemically modified enzymes in terms of activation or inactivation is somewhat artificial. For example phosphorylation leads to an activation of glycogen phosphorylase but an inactivation of glycogen synthetase. These chemical modifications could equally well be considered as dephosphorylation in which case glycogen phosphorylase is inactivated and glycogen synthetase activated. However, from an operational point of view, whether an enzyme is activated or inactivated is of paramount importance.

Enzyme Activation

Four enzymes of the Calvin cycle and two enzymes of the Hatch-Slack pathway are light activated whilst glucose-6-phosphate dehydrogenase, an enzyme of the pentose phosphate pathway is light inactivated.

It is highly probable that a number of these responses reflect a reductive process, since dithiothreitol treatment of crude extracts gives results similar to light treatment of intact seedlings. However, in some cases the reaction is certainly more complicated than a direct response to a change in the SH/SS redox potential.

The mechanism of activation may vary from species to species. Thus the light-activation of pea NADP glyceraldehyde phosphate dehydrogenase appears to involve a conformational change produced by reduction and leads to an increase in V_{max}. The spinach enzyme however, appears to need NADPH or NADP or

ATP in addition to a reducing agent for activation. The activation involves dissociation of a tetramer leading to an increased affinity for NADP.

Figure: Schematic representation of the processes of activation and inactivation of pyruvate, P: dikinase. The process leading to the irreversible inactivation of both active and inactive enzyme* in vitro *apparently include an O_2-dependent oxidative reaction involving thiol groups. (After Hatch & Slack, 1969.)

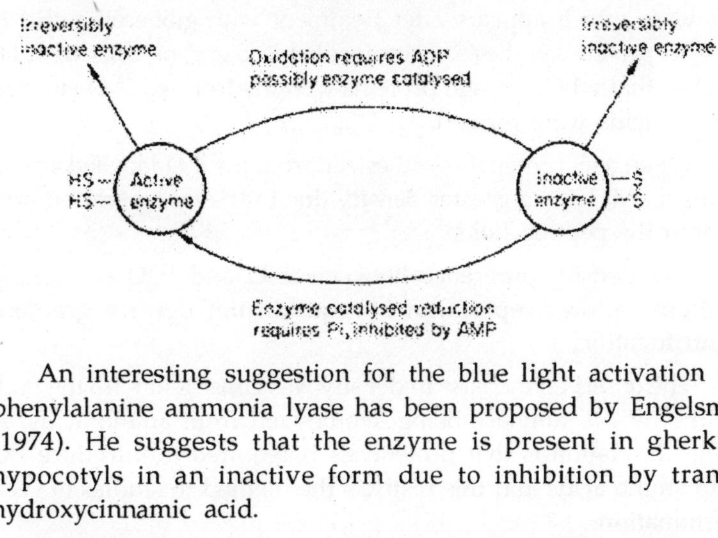

An interesting suggestion for the blue light activation of phenylalanine ammonia lyase has been proposed by Engelsma (1974). He suggests that the enzyme is present in gherkin hypocotyls in an inactive form due to inhibition by trans-hydroxycinnamic acid.

Blue light (in the presence of a photoreceptor such as riboflavin) converts the trans- to the cis-hydroxycinnamic acid which is much less inhibitory. A phytochrome-mediated activation of phenylalanine ammonia lyase has also been reported in mustard cotyledons by Attridge *et al.,* (1974). It is not clear if the cis-trans isomerization is applicable to the phytochrome-mediated activation which is inhibited by cycloheximide.

Enzyme Synthesis

In the absence of activation or inactivation, the net synthesis of an enzyme is given by gross synthesis minus degradation. Net synthesis is measured as the change of activity so that gross synthesis can be measured if the rate of degradation is known. Because of the technical difficulties involved in obtaining quantitative data most workers have been concerned to develop methods which establish that the increase in activity of an enzyme is due to synthesis rather than degradation.

Density Labelling

Gibberellic acid stimulates the synthesis of a number of enzymes in the aleurone layer of barley seeds. Filner and Varner (1967) introduced a method of demonstrating that the bulk of α -amylase which appears after treatment with gibberellic acid is due to protein synthesis. They incubated aleurone cells with $H_2\,^{18}O$ so that when storage proteins were hydrolysed ^{18}O-labelled amino acids were formed.

When new protein is synthesized from the ^{18}O-labelled amino acids it will have a greater density due to the incorporation of ^{18}O into the peptide links.

The density difference between ^{18}O and ^{16}O containing proteins allows separation by equilibrium density gradient centrifugation.

The observed increase in density was consistent with the bulk of the new α -amylase being synthesized from amino acids. A major limitation is that proteolysis is required to introduce ^{18}O into amino acids and this restricts the method to studies of seed germination.

An alternative method is to introduce ^2H into the protein. This can be achieved by incubating a tissue in ^2H_2O so that ^2H labels the amino acids by entering at the α -carbon atom in a transaminase catlysed exchange reaction. With prolonged incubation in ^2H_2O, deuterium enters the amino acids at other points producing large changes in the density of newly formed protein.

The major advantage of the density labelling methods is that they enable the synthesis of a particular enzyme to be demonstrated without purifying the enzyme. However, the method requires expensive equipment and is time consuming. Consequently many investigators rely on data obtained by the use of inhibitors of protein synthesis. Such data is rapidly obtained and at low cost but its interpretation must always be equivocal.

Enzyme Induction

Filner *et al.*, (1969) have listed the numerous examples of increases in enzyme activity in response to age, hormonal or environmental changes. In many cases these increases may be due to enzyme synthesis, though the evidence is often little more than that gained by the use of inhibitors of protein synthesis.

The evidence for substrate-induced enzyme synthesis is extensive in the case of nitrate reductase but in other cases, for example the induction of a specific isoenzyme of glutamate dehydrogenase, the case is not yet proven.

FINAL COMMENTS

The basic similarity of living organisms is reflected in their biochemistry, the major metabolic pathways being common to all living organisms.

The diversity of living organisms is reflected in their specialized control systems. The control mechanisms of one species may be very different from those of a closely related species.

The investigator has a daunting task when he considers the amount of work necessary to establish control mechanisms over a wide range of species. The difficulties of comprehending control mechanisms even in a single organism are no less daunting.

The difficulties stem from the number of assays necessary to establish the complete rate law for a regulatory enzyme. For example the enzyme glutamine synthetase from *E. coli* is known to be affected by eight reactants and modifiers. Hence if 6 points per curve are necessary for a kinetic analysis, then to establish the rate law for glutamine synthetase will require in excess of 1.6×10^6 assays.

The task of analysing the many enzymes involved in metabolism is frightening and when we recognize that all the enzymes interact and affect one another the system defies analysis even with modern computers.

The way ahead seems to require inspired guesses at the nature of the control mechanisms involved, followed by model building and a systems analysis—probably making linear approximations—to see if the behaviour of the biological system can be predicted.

The difficulty in analysing control systems may be compared with the difficulties in understanding economics. The main control step of glycolysis appears to be phosphofructokinase, a reaction which involves a loss of free energy of 4 Kcals per mole. This loss may be considered the cost of controlling glycolysis. Protein degradation involves a large loss of free energy and this may be the price the cell has to pay for its complicated but effective control mechanisms. The situation can be likened to the economy of surplus production in society—in the Brave New World of Aldous Huxley products were destroyed to maintain the economy—perhaps nature is a capitalist!

HORMONE ACTION IN PLANTS

The genesis of hormone research in plants occurred with the discoveries of Charles Darwin and his son Francis, who published, almost a century ago, their book entitled The Power of Movement in Plants'. Their experiments dealing with phototropism revealed that an 'influence' produced in the tip of a canary grass coleoptile was translocated to the basal portion where it caused the organ to grow towards the light. Almost fifty years later Went isolated a substance with just these properties which was produced in the tip of coleoptiles and transported to the basal portion of the organ. This substance was named auxin. Ten years after Went's (1934) discovery, the structure of this material was identified as indoleacetic acid (IAA).

The nature of auxin effects on plant growth and development was studied in many laboratories in the late 1930's and 1940's. However, it was not until World War II that the selective action of indoleacetic acid served as a basis for developing new chemicals for the control of growth in specific plants. At that time phenoxyacetic acids were discovered and their derivatives were carefully studied under the cover of military secrecy, both in the United States and England. The results of those studies were first published in scientific journals in 1945 and 1947. Today the most widespread uses of growth regulators are as herbicides.

Plant hormones may be divided into five general groups. Auxins include the native indoleacetic acid, as well as the synthetic phenoxyacetic acids, notably 2,4-D (2,4-dichlorophenoxyacetic acid). The second group is the gibberellins which are steroid-like compounds comprising over forty different structures. The third group, the cytokinins, are all N^6 substituted adenosine compounds. The fourth class is represented by a single substance, abscisic acid, which is also derived from isoprenoid units as are the gibberellins. Ethylene, the simplest of all the plant hormones, is a gas and therefore is easily spread from organs of production to organs of sensitivity.

It is interesting that of these five classes of hormones, two are generated from mevalonic acid. Ethylene is produced by the metabolism of methionine and indoleacetic acid is produced by the removal of carbon and nitrogen from tryptophane. Cytokinins, in terms of synthesis, are probably the most complex hormones. They are possibly produced from the breakdown of transfer RNA (tRNA). In the intact tRNA, the adenosine moiety of the RNA is modified by the addition of an isoprenoid group; subsequently, the tRNA is probably degraded, yielding isopentenyl adenosine, (IPA) the endogenous cytokinin.

One might reasonably imagine that an understanding of the physiological role and biochemical mechanisms of action of the plant hormones would have been elucidated quite some time ago. This of course, is in relation to what is known about the many and complex hormones in animals. Unfortunately, however, the plant hormones are not as well understood. Probably the main obstacle to the elucidation of the action of plant hormones has been the fact that they have overlapping and complementary effects on the actions of each other.

During the years between 1960 and 1970 substantial progress appeared to be made on the biochemical mechanisms of action of plant hormones. Much of this research dealt with the effect of the hormones on nucleic acid metabolism, protein synthesis and the synthesis of specific enzymes, since it was felt that the mechanism of hormone action would involve the control of the production of messenger-RNA (mRNA) and the *de novo* synthesis of enzymes. In the late 1960's the general nature of the research was radically changed because many investigators had found that the plant hormones mediated an effect on cell growth within a few minutes. This information implied a very rapid primary response mechanism of hormone action; consequently, these effects could not be mediated by changing the rate or type of synthesis of nucleic acids or of specific proteins. Many years of research have since been devoted to studies dealing with short-time growth responses to plant hormones. It now seems most likely that plant hormones regulate plant growth through both short-term and long-term growth controls.

It is fairly well understood that animal hormones first interact with the animal cell by binding to, or reacting with, a receptor site of some type within the target cell. Plant physiologists and biochemists are now beginning to consider that the action of many of the plant hormones similarly involves their binding to receptor agents which in turn amplify the action of the hormone and thereby bring about specific changes in nucleic acid synthesis, protein synthesis, enzyme activity and possibly other physiological responses, such as changes in permeability of membranes.

This chapter will summarize the historical evidence which has led to the current level of understanding of the action of plant hormones. Emphasis is placed on the reactive sites within target cells and how the interaction of the hormones with their specific receptors may regulate the growth and development of plant cells.

When a hormone is applied to a responsive plant system it brings about a specific change which results eventually in a measurable biochemical or physiological effect. Two distinct aspects of the measured effect are involved: the specific change in metabolism and the series of steps which lead to the physiological effect. Usually, the molecular interaction of the hormone at its site of action is referred to as the *mechanism* of action. The subsequent sequence of reactions leading to the physiological effect is referred to as the *mode* of action. Therefore, by definition each hormone has its own distinctive mechanism of action even though the manifestation of the hormone mechanism may depend upon prior action by other factors. Thus, it is possible for the hormonal *mechanisms* in one plant system to lead to a series of physiological responses which may be completely different from those in a second system. This difference in *mode* of action can be brought about because the second system has more or less of another hormone, has other biochemically rate-limiting components or possesses structural and cytological differences.

AUXIN ACTIONS

Amongst plant growth regulators, the auxins have been studied the longest and have involved the greatest amount of research. This class of hormones is known to promote cell enlargement or cell elongation, a process which requires extension of the cell wall.

Commencing in 1940, the majority of studies involving auxins were begun in an attempt to determine the molecular interaction with the plant cell. Most of this research was directed toward the cell wall because it was believed that auxin action required a change in the cell wall to allow for cellular expansion. Thus, the effects of the hormone were thought either to change the cell wall deposition or to hydrolyse certain cross-linkages in cell walls making them more elastic.

This type of research was carried out for several years without achieving a clear understanding of how auxin might control such processes within the cell wall. In the 1950's a new area of research was begun. Under the direction of F. K. Skoog in Madison, Wisconsin, a report (Silberger & Skoog, 1953) was published which showed that the auxin, indole-3-acetic acid, remarkedly affected the RNA and DNA contents of plants.

Auxin increased the content of nucleic acids in tobacco tissue cultured on a sucrose-agar medium. This increase occurred prior to the auxin-induced growth of the tissue at concentrations of IAA which were optimal for cell enlargement. For many years following this discovery much research was devoted to the general mechanism by which auxin increased the synthesis of nucleic acids.

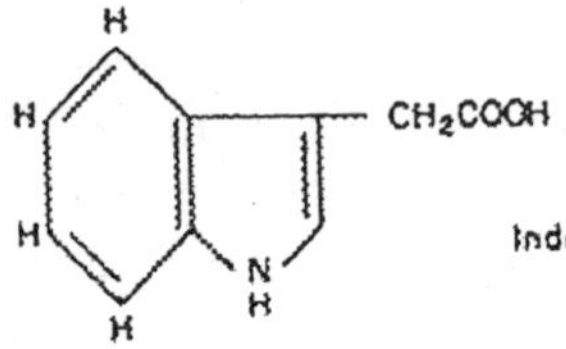

Indoleacetic acid (IAA)

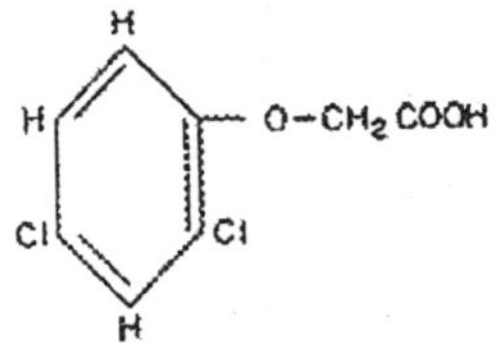

2,4 - Dichlorophenoxyacetic acid (2,4 - D)

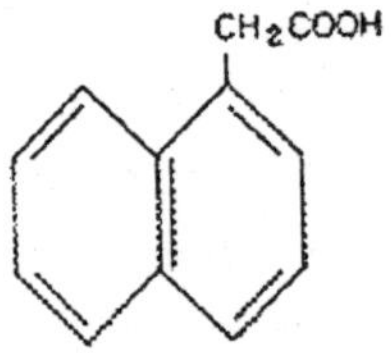

Naphthaleneacetic acid (NAA)

Figure: Structure of various auxins.

A great share of this research was begun in J. Hanson's laboratory in Urbana, Illinois by West and Key (West *et al.*, 1960) who showed that the synthetic auxin, 2,4-D, produced a wide range of morphological and physiological changes in the hypocotyl of soybean and the mesocotyl of corn.

Within 15–24 hours after 2,4-D treatment, cellular enlargement was noted concomitant with an increase in the size of the nucleus. Accompanying these changes were dramatic increases in RNA content, most of which were due to increased ribosomal-RNA. Chrispeels and Hanson (1962) suggested that auxin acts on the nucleus causing it to revert to a meristematic type of metabolism. The role of the nucleus in such a sequence of events is of obvious importance since the nucleus has been shown to accumulate RNA in response to auxin. Beginning in the 1960's, Key began experiments on auxin regulation of nucleic acid synthesis. He and

Ingle (1964) demonstrated that auxin controls the synthesis of nucleic acids other than that of ribosomal-RNA. Their data revealed that auxin causes production of nucleic acid which appears to be of the messenger-RNA type.

Subsequent experiments by O'Brien *et al.*, (1968) showed that treatment of soybean with auxin caused a large increase in chromatin-directed RNA synthesis. It was of interest to note that auxin-induced RNA synthesis produced a type of RNA which was different from the control RNA as judged by molecular size and nearest neighbour analysis. Following this area of research other studies (Hardin & Cherry, 1972; Hardin *et al.*, 1970; Hardin *et al.*, 1972) showed that 2,4-D increased the activity of RNA polymerase. Current research suggests that particular cytoplasmic or membrane bound factors may enhance the activity of RNA polymerase. It is hypothesized (Hardin *et al.*, 1972) that the mechanism of action of auxin is to bind to a receptor molecule located on the plasma membrane. The receptor is then released and moves into the nucleus where it modulates the activity of the RNA polymerase. The increase in RNA polymerase activity leads to increased synthesis of messenger-RNAs, which in turn regulate, or control, the synthesis of specific proteins within the target cells.

A primary challenge is to isolate the factor which binds to, or reacts with, auxin, and determine whether this is the primary action of the hormone. Secondarily, research is needed to determine what series of events this interaction puts into metabolic play. It is likely, once the auxin has bound to the receptor molecule within the target cells, that the changes in nucleic acid synthesis cause an increased activity of the various enzymes associated with the cell wall. Many other activities which have been measured over the many years are probably secondary effects resulting from the primary action of the auxin reacting with its receptor molecule.

Regulation of Cell Wall Extension Ability

When plant cell walls elongate, an accompanying increase in weight and volume also takes place. This increase in size requires that the cell wall be increased in mass as well as area. Even though the dry weight of wall material greatly increases due to cellular enlargement, the thickness and density of the wall remains constant. Thus, wall synthesis appears to be a fundamental requirement during cell elongation. In addition, an increase in wall mass must come from a deposition of polysaccharides. It was believed for many years that auxin, in this particular case indoleacetic acid, increased cell elongation by promoting cell wall synthesis. Auxin, in order to allow for the increased wall extensibility, may also regulate the activity of enzymes involved in cell wall loosening which catalyse the breaking of various cross-linkages between the wall microfibrils.

Plant tissues which respond to auxin by an increased rate of cellular elongation also exhibit an increase in cell wall loosening. However, the kinetics of auxin-induced wall loosening vary considerably from tissue to tissue. For example, in dark grown maize coleoptiles wall loosening is induced by concentrations of auxin around 10^{-4} M, and the increase in loosening proceeds very slowly with time. However, the same tissue responds to auxin very dramatically as maximal increases in total length occur between two and six hours after exposure to auxin. It is of interest, therefore, that the rate of cell wall elongation, as measured by increased section length, is much greater than the increase in cell wall extensibility.

Thus, considerable changes in cell wall extensibility in a given tissue as caused by auxin, may affect growth in a different manner. In certain tissues the rate of growth is constant whilst total wall extensibility changes. In other tissues wall extensibility remains constant during a period of changing growth rate. At the present time it seems that auxin causes a biochemical change in the wall probably by breaking or modifying the cross-links between the wall polysaccharide chains. These changes in the cell wall are then translated under cell turgor pressure into wall elongation. It is now relevant to discuss how auxin affects the cross-links in a cell wall and what the biochemical changes are which take place in a cell wall during cell elongation. In this regard, it is necessary to look at the effect of auxin on various enzymes associated with the cell wall in the context of how these enzymes affect cell wall loosening properties.

Action on Cell Wall Associated Enzymes

The most dramatic auxin effect on enzyme formation is the induction of cellulase, polygalacturonase and other hydrolases in pea cotyledons treated with a high concentration of IAA (Datko & Maclachlan, 1968). Detection of *in vitro* formation of cellulase by a ribosomal preparation from peas has been claimed. This was reported to be enhanced using a ribosomal system prepared from IAA-treated peas (Davis & Maclachlan, 1969). The large IAA effect on cellulase formation *in vivo* is a slow response occurring over several days and cannot be involved in a rapid auxin action on elongation. However, it could be a cause of the lateral swelling response seen in pea cells following IAA treatment.

Auxin stimulates not only the synthesis of cellulase but also the level of glucan synthetase. Masuda (1968) and Masuda and Yamamoto (1970) found that the fungal *b* -1,3-glucanase, isolated from cultures of *Sclerotinia libertans* induced rapid elongation of excised oat coleoptile segments. This enzyme was shown to increase cell wall extensibility as measured by a stretching method. In other studies, Masuda *et al.* (1970), compared the activity of the endo-*b* -1,3-glucanase and its activity on *Avena* coleoptile cells with those of the exoenzyme. They found that exoglucanase enhanced elongation and extensibility of the cell wall. But the effect was not additive to the effect of IAA. Furthermore, at least three hours of incubation with exoglucanase was required for enhancement of elongation. The endoglucanase showed no effect on cell wall elongation. Cell wall turnover and auxin effects thereon in pea stem tissue have been studied using pulse-chase wall labelling experiments. Considerable turnover of galactan occurs but this is not influenced by IAA (Labavitch & Ray, 1974).

The action of auxin on soluble xyloglucan can be assayed with relative ease and precision and has been positively demonstrated to be in progress within thirty minutes after exposure to IAA and probably within fifteen minutes, placing it amongst the most rapid metabolic effects of auxin. This auxin effect is blocked by metabolic inhibitors that are known to block elongation, but persists under complete osmotic inhibition of elongation by mannitol. It seems likely that it is involved in the action of the cell wall that leads to elongation in pea cells. Involvement of xyloglucan is understandable in terms of a recent model for cell wall structure, according to which xyloglucan serves to bind matrix polysaccharides to cellulose microfibrils.

Hydrogen-Ion Pump

Since the 1930'S it had been known that acidic media can stimulate elongation in auxin-sensitive tissue. Recent work showed that acidic buffers or CO_2 solutions in the pH range from 3 to 4 induce a rate of coleoptile cell elongation as great as or greater than that obtainable with auxin (Rayle & Cleland, 1970; Hager *et al.*, 1971; Evans *et al.*, 1971). Acid-induced elongation starts almost immediately upon treatment rather than after the 10–15 minute latent period characteristic of auxin action.

Acid-induced elongation is not suppressible by metabolic inhibitors, such as cyanide, mercurials, and cycloheximide, or by lack of oxygen, all of which block auxin-induced growth, and appears to be a passive process independent of metabolism. In later work it was found that when the epidermis is removed to improve H^+ entry into the tissue, the full acid-pH stimulation of elongation develops over the pH range from about 6.0 to about 5.0 (Rayle, 1973; Cleland, 1973).

Rayle *et al.*, (1970b) found that cell wall skeletons of frozen-thawed, dead coleoptile tissue would elongate dramatically if treated with acidic media while being held under tension by an applied load. Rayle and Cleland (1972) concluded that acid pH- and IAA-induced elongation must occur by the same mechanism, because the two kinds of elongation had similar rate, similar temperature dependences, and a similar yield threshold. While suggestive, these similarities do not establish identity of biochemical mechanism.

Hager *et al.*, (1971) also felt, on grounds that IAA-induced elongation could be suppressed by alkaline media, that IAA-induced growth involves the same biochemical mechanism as acid-induced elongation. In support of their proposed auxin-stimulated H^+ pump they offered experiments showing rapid stimulation of coleoptile elongation by ATP, ITP, and GTP under anaerobic conditions, although these results may have represented merely the acid pH effect itself, since these compounds are strong acids and were applied in unbuffered, pH 5 media.

In recent work an auxin-induced release of H^+ ions from coleoptiles stripped of their epidermis (Cleland, 1973; Rayle, 1973) and from other auxin-sensitive tissues (Marrè *et al.*, 1973; Ilan, 1973) has been detected. This is measured simply as a gradual fall in pH of the medium bathing the tissue, upon treatment with IAA, from near-neutral pH to pH 5 or below, i.e., into the range that by itself induces elongation. Detectable H^+ secretion by coleoptiles begins within 20 to 30 minutes after exposure to IAA and may be regarded as a 'rapid response'. To the extent so far studied, H^+ secretion by coleoptiles has a specificity for auxin analogues and antagonists similar to the specificity seen in growth, and a similar dependence on metabolism. Secretion is sensitive to inhibitors and uncouplers of energy metabolism and, perhaps unexpectedly for a transport process but just like cell enlargement, H^+ secretion is quickly inhibited by low concentrations of cycloheximide (Rayle, 1973; Cleland, 1973). Comparable results with pea stem segments have been reported by Marrè *et al.*, (1973).

These findings, coupled with the observations that induction of elongation by auxin is prevented by sufficiently well-buffered media of pH 6 to 8 provided the cuticle is removed or rendered permeable by gentle abrasion (Rayle, 1973), constitute presumptive evidence that the auxin effect on cell enlargement is mediated by externally secreted H^+ ions.

The simplicity and directness of the acid secretion theory of auxin action is appealing and the theory has quickly become popular. Various auxin effects on cation and anion transport noted previously were considered by their authors to be consistent with the H^+ secretion theory.

H^+ secretion could involve either a parallel flow of anions, or a counterflow of, or exchange with, cations as has been inferred for H^+ pumping by beet cells (Poole, 1973). However, as yet no immediate dependence on external ions of either auxin-induced H^+ secretion or growth has been found.

Cleland (1973) and Rayle (1973) observed that auxin-treated coleoptiles cease to release H^+ when the medium reaches pH 5.0 or slightly below, suggesting that auxin-stimulated H^+ secretion is inhibited by acid pH, like other H^+ pumps (Poole, 1973), or alternatively that, at pH 5, back-diffusion of H^+ into the cell offsets the action of the pump. This feature should allow extensionpromoting pH values in the cell wall to be reached rapidly under auxin treatment, while shutting down further acidification which would probably be injurious to the cell (Cleland, 1973).

Some objections to the H^+ secretion theory of auxin action and weaknesses in the evidence for it should be mentioned. For example, it has not yet been shown that suppression of auxin elongation by neutral buffers is not due to inhibited uptake of auxin, or to failure of primary action (the auxin binding found by Hertel *et al.,* (1972) had a pH optimum below pH 6.5).

The most important piece of evidence still needed for the acid secretion theory is that the pH in the tissue's free space falls to an extension-stimulating value at the time that rapid auxin-induced elongation commences, i.e., within 10–15 minutes after exposure to auxin, as against the hour or more that is required for an external bathing medium to reach pH 5.0 to 5.5 (Cleland, 1973; Rayle, 1973). At the very least it must be demonstrated that IAA induces H^+ secretion as quickly as it induces rapid elongation. Existing data fail this test and show an increase in H^+ efflux from coleoptiles beginning, at the earliest, 20 or 30 minutes after exposure to IAA (Cleland, 1973; Rayle, 1973). These authors feel that the discrepancy may be attributed to time lags imposed by the free space volume and the diffusion path length for H^+ to reach the external medium. A worse timing discrepancy was seen with sunflower hypocotyl (Ilan, 1973; cf. Uhrström, 1969). A more sensitive method for measuring H^+ efflux needs to be employed to resolve these discrepancies. Better still, the pH within the free space of the tissue should be measured directly during response to auxin treatment.

Regulation of Genetic Material

After the discovery of messenger-RNA it was suggested that the regulation of gene action by auxin required the specific control of messenger-RNA synthesis. Using a series of inhibitors, Key (1964) and Key and Ingle (1964) noted that treatment of soybean hypocotyl with 5-fluorouracil (5-FU) inhibited total RNA synthesis by 80% whilst growth proceeded normally. Further studies on the fractionation of nucleic acid from soybean hypocotyls on a methylated albumin Kieselguhr column showed that 5-FU inhibited the incorporation of labelled adenosine diphosphate into ribosomal-RNA but had little effect on the labelling of messenger-RNA. Additional experiments with soybeans showed that 5-FU did not affect the tenaciously bound RNA whose synthesis is promoted by 2,4-D.

Sen and his colleagues (Roychoudury & Sen, 1964; Roychoudury *et al.*, 1965) showed that nuclei isolated from coconut milk responded to auxin by making more RNA *in vitro.* On the basis of these observations, it was proposed that auxin directly affected the nucleus and thereby regulated gene expression. It was thought that in the presence of the hormone a greater amount of DNA template was exposed and allowed to be transcribed into RNA. In an attempt to test this hypothesis, Cherry (1967) showed that nuclei isolated from peanut cotyledons did not respond *in vitro* to 2,4-D at physiological concentrations (10^{-8} to 10^{-6} M). From many other experiments, Cherry demonstrated that only in approximately 1 out of 10 attempts could nuclei be shown to respond to the hormone *in vitro* by the synthesis of more RNA. In general, absolutely no effect of the auxin on the capacity of the isolated nuclei to produce RNA was found. On the other hand, it was found routinely that isolated nuclei from soybean seedlings pretreated with 2,4-D produced twice as much nucleic acid as did control nuclei.

Furthermore, experiments of O'Brien *et al.*, (1968) showed that chromatin isolated from soybean hypocotyls did not respond *in vitro* to 2,4-D. In all cases the tissues needed to be treated with the growth regulator for at least two hours before any effect on chromatin-directed RNA synthesis could be noted. It is to be noted, however, that in a few experiments an effect of the hormone could be observed within thirty minutes. Subsequently, a progressive increase in chromatin-directed RNA synthesis was noted as a function of time. From those experiments, it was suggested that the auxin-enhancement was a result either of a more active RNA polymerase or of gene derepression leading to an increased availability of DNA template. A third possibility was that both of these effects were involved. Subsequent experiments demonstrated that in the presence of *E. coli* RNA polymerase, chromatin from both control and 2,4-D treated plants allowed the synthesis of similar amounts of RNA at saturation with the polymerase. Even though the total *E. coli* RNA polymerase activity with chromatin from 2,4-D treated plants was slightly higher than that with control chromatin, it was concluded that the auxin primarily promoted the endogenous RNA polymerase, rather than the chromatin template availability.

When it became possible to solubilize RNA polymerase from chromatin (Hardin & Cherry, 1972) two important aspects were noted. First of all, it has not been possible to show that the addition of auxin *in vitro* to solubilized RNA polymerase, or to chromatin, increases the rate of RNA synthesis (Hardin *et al.*, 1970). Secondly, of the many experiments that have been performed, it appears that treatment of sensitive plants with auxin leads to a greater production of RNA polymerase I (Hardin & Cherry, 1972; Hardin *et al.*, 1972). This is the enzyme which is thought to be present in the nucleoli, as judged from comparisons with animal RNA polymerase, and is the enzyme thought to transcribe ribosomal into ribosomal-RNA. These data agree with the fact that large increases in ribosomal-RNA and ribosomes are observed after auxin treatment. However, they are inconsistent with the idea that auxin increases the activity of the RNA polymerase enzyme which transcribes unique DNA sequences into messenger-RNA.

Action on Membranes

A current, interesting view of auxin action is that the hormone rapidly effects its action at, or on, cellular membranes, possibly involving the regulation of the export of growth active materials across the plasma membrane into the cell wall space. Auxin action at the plasma membrane was hypothesized by Hertel and Flory (1968) and by Rayle *et al.* (1970) partly in the belief that auxin transport and auxin action on growth involve closely related, if not identical, interactions, presumably with some carrier site located in the plasma membrane. Evidence for auxin action at the plasma membrane comes from observed IAA effects on membrane potentials (Tanada, 1970). This effect, however, has not yet been adequately studied nor shown to be related to the elongation process. Another indication of auxin action at the plasma membrane is the induction of cell elongation by certain fungal toxins which are known to increase membrane permeability (Evans, 1973).

It is assumed that specific auxin receptor sites must be located in or on the plasma membrane if the hormone is to change the permeability or to change any physical property within the membrane. With this general assumption in mind, Lembi *et al.* (1971) began to look at the binding of labelled NPA (naphthylphthallamic acid) a powerful competitive inhibitor of polar auxin transport which was considered to be likely to bind to specific auxin binding sites.

Using maize particle preparations, they tentatively showed that there was a greater amount of NPA binding to fractions rich in plasma membranes in comparison to all other fractions obtained. IAA did not compete with NPA for the sites which thus do not seem to be the actual carrier sites for polar transport of IAA. It is of course conceivable that the sites binding NPA are not the same sites which would normally bind the auxin, IAA.

Hertel *et al.* (1972) detected a specific binding of labelled IAA and naphthaleneacetic acid (NAA) to certain particles in maize homogenates. Specific binding was assayed as radioactivity retained by particles exposed to 2 to 10 × 10^{-7} M labelled auxin minus radioactivity retained when, in addition, 10^{-4} M unlabelled IAA or NAA was added to compete with and displace labelled auxin from saturateable binding sites. The radioactivity which remained in the latter case then was termed unspecific binding.

This difference-assay gave specific binding with binding constants in the rage of 10^{-} ; to $10^{-}{}_{5}$ M. Specificity of the binding as an auxin phenomenon was indicated by the evidence that analogues of IAA or NAA which are active on growth, and are transportable, can displace labelled IAA or NAA from the binding sites, whereas chemically similar, but biologically inactive, molecules do not compete in the binding assay.

One defect of these data was that 2,4-D, a very active auxin, competed relatively weakly with IAA or NAA in the assay, nor was specific binding of labelled 2,4-D itself detected. Based on the evidence that auxin-specific binding sites are localized in plasma membranes, Hardin *et al.* (1972) proposed that the reaction of auxin, either the native IAA or the synthetic 2,4-D, with the plasma membrane would probably lead to conformational changes within the membrane, causing the release of a receptor molecule into the cytoplasm. This receptor molecule may or may not be the same binding agent with which the auxin interacts. Nevertheless, the receptor released from the plasma membrane could then travel through the cytoplasm and into the nucleus where we know auxin ultimately increases the activity of RNA polymerase.

From previously published information from animal systems, (Shyamala & Gorski, 1969; Jensen *et al.*, 1968) a good analogy can be drawn with hormone interaction with receptor molecules which then migrate into the nucleus where nucleic acid synthesis is controlled. Following these lines of investigation, Hardin *et al.* (1972) isolated a plasma membrane-rich fraction in which the plasma membrane was thought to be 70% homogeneous. The addition of plasma membranes to soybean RNA polymerase caused an increase in RNA synthesis *in vitro*. Subsequently, it was shown that pretreatment of the membranes with 10^{-7} M 2,4-D greatly increased RNA polymerase activity. Still further, when the plasma membranes were pretreated with 2,4-D followed by removal by centrifugation, the supernatant fraction contained the RNA polymerase stimulus. The chemical release of this factor appears to be specific for auxin, i.e., IAA and 2,4-D bring about this release, but a non-auxin such as 2,5-D is totally inactive.

In other studies an attempt was made to determine which of the multiple RNA polymerases were stimulated by the auxin-released factor. It was found that the addition of plasma membrane factor to a semi-purified RNA polymerase preparation enhanced the *a* -amanitin sensitive polymerase. By analogy with animal RNA-polymerase studies, these data imply that auxin increases the nucleoplasmic RNA poly merase, the enzyme which is thought to transcribe DNA into messenger-RNA. If this is true then it is sensible to believe that auxin, through the release of a receptor molecule, may regulate the transcription of unique DNA sequences through the control of a specific enzyme.

SUMMARY

Because of rapid auxin responses on plant cells, over the last few years a whole new field of research has begun which relates not only to the gene concept but also to rapid responses. Therefore, the new area of research has dealt with actions of auxins on receptors, membranes and other binding surfaces which could lead to a rapid growth response, as well as a continued and long response involving nucleic acid synthesis and protein synthesis. The apparent binding of auxin to plasma membrane sites has stimulated interest in the concept that the hormone might bring about conformational changes within the membrane structure. This in turn might lead to changes in permeability, causing an increased ion flux across the membrane. If a receptor molecule is located on the plasma membrane, a change in conformation of the membrane might bring about a release of the receptor. It has been proposed that this receptor moves into the nucleus of the target cell and once there, increases the activity of RNA polymerase. This, in turn, leads to the synthesis of messenger-RNA which codes for proteins and this brings about the net result of auxin increased growth.

In this particular section the effects of cyclic-AMP on auxin controlled growth have not been covered. When cyclic-AMP was first found to stimulate plant growth, considerable excitement was generated (Solomon & Mascarenhas, 1971). However, that excitement was short-lived, and it now appears that cyclic-AMP has little or no effect on plant cells (Ownby *et al.*, 1975).

A number of enzymes have been shown to be affected by auxin. These include the hydrolases, particularly cellulase, and other enzymes associated with cell wall degradation as well as cell wall synthesis. Whilst these enzymes are affected by the auxins, it is very likely that this result is a secondary effect of the auxin and is not related to the primary mechanism of action. In essence then, it appears that auxins control a multitude of physiological and biochemical responses in plants. It is possible that the primary site of action is localized within the plasma membrane or some surface binding site of the cell. A change at this particular level, either in structure or function, could bring about the release of a factor which moves into the nucleus. As a secondary amplification, the hormone thus modulates the activity of RNA polymerase. This in turn leads to the synthesis of RNA from parts of the genome which were either not transcribed at all, or just to a low level in the absence of the receptor. A change in DNA transcription in this manner would lead to the synthesis of new proteins which would bring about the change in the growth potential.

Figure: A hypothetical model of auxin action. The interaction of auxin with the plasma membrane results in the release of a factor which moves through the cytoplasm and into the nucleus. The factor controls the activity of RNA polymerase II in the nuclei which leads to synthesis of mRNA. The new species of mRNA translated in the cytoplasm into new proteins leads to enhanced cellular growth.

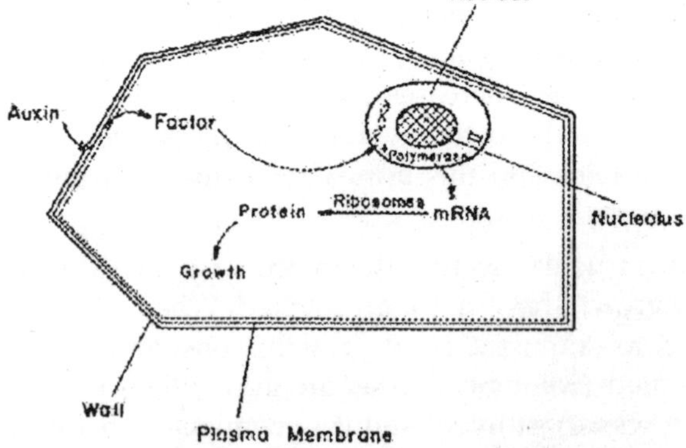

GIBBERELLIN ACTION

Gibberellins may be defined as compounds having a gibbane skeleton and biological activity in stimulating cell division or cell elongation or both.

Gibberelins also may possess other biological activities such as induction of nucleic acid and enzyme synthesis. Presently there are more than forty known gibberellins which have been isolated from green plants or the fungus *Gibberella fujikuroi* (Cherry, 1973). Since the late 1960's three to four new gibberellins have been identified each year.

Thus, it seems that great progress is being made in the identification of naturally-occurring gibberellins; on the other hand the mechanism of action of gibberellins is still unknown. Plant physiologists are therefore confronted with a very wide array of gibberellins, several of which often occur in the same tissue, and the question arises as to their function within the cells.

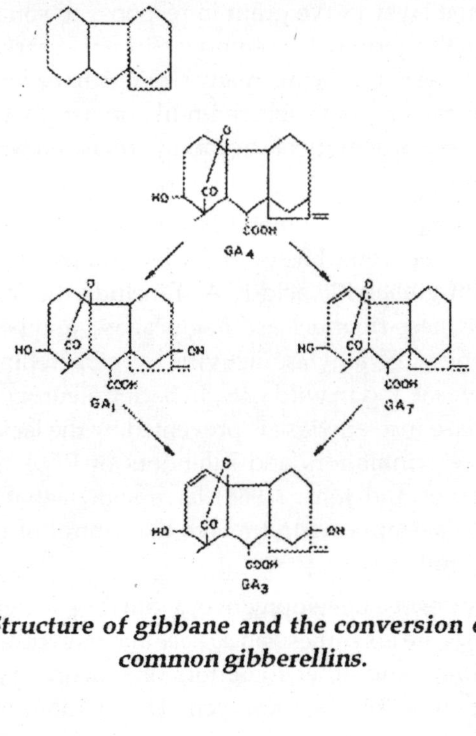

Figure: Structure of gibbane and the conversion of GA$_4$ to common gibberellins.

Work beginning in the 1960's on the possible regulation of gene transcription by gibberellins was initially successful and achieved wide-spread acceptance.

Later work raised many doubts, however, and at the present time, many workers in the field reject the concept that gibberellins regulate physiological processes through the control of gene transcription. Alternative mechanisms involving an action of gibberellins on lipids and membranes are now being proposed.

Gibberellin Control of Enzyme Synthesis

The aleurone layer of cereal grain is composed of a relatively homogeneous population of non-dividing cells which respond as a target tissue to gibberellins in the concentration range of 10^{-7} to 10^{-10} M (Varner & Johri, 1968). It is, therefore, an ideal tissue for a study of the hormonal control of enzyme activity. Haberlandt in 1890 first discovered the secretion of starch liquifying enzymes by the aleurone layer in rye grain in response to some substance obtained from the germinating embryo. Since that early discovery, over the many years following, many workers have found embryo substances or factors and gibberellin-like materials which bring about an increase in activity of several hydrolase enzymes (Varner & Johri, 1968).

Some of the enzymes which have been shown to increase in activity in the aleurone layers of barley grains in response to treatment with gibberellic acid (GA_3) include the following: *a* -amylase, proteinase ribonuclease, *b* -glucanase, and pentosanases. A similar control of amylase activity by gibberellins has been observed in wheat and in wild oats. In barley aleurone layers GA-induced increase in *a* -amylase is prevented by the lack of oxygen, by cytochrome *c* inhibitors and inhibitors of RNA and protein synthesis. Varner and Johri (1968) have shown that gibberellin control of amylase appears to involve the control of nucleic acid and protein synthesis.

GA_3 enhances the development of *a* -amylase activity in cereal grains. The enzyme is synthesized *de novo* and its synthesis blocked by cycloheximide and other inhibitors of protein synthesis. It is of interest to note that actinomycin D (an inhibitor of RNA synthesis) only inhibited the production of *a* -amylase when the tissue was exposed to the inhibitor early in the incubation process. If, however, the tissue was exposed to actinomycin D at some point after enzyme synthesis had begun, little or no inhibition occurred. Thus, it was not certain whether the hormone actually controlled the synthesis of RNA at the level of transcription or controlled the translation of pre-existing messenger-RNA. Many attempts were initiated to answer this question; unfortunately, to date, a definitive decision has not been possible. The most likely conclusion is that the hormone does not control the synthesis of *a* -amylase at transcription (Chrispeels & Varner, 1967).

The Effects on RNA Synthesis

Considerable indirect evidence has shown that GA_3 influences the synthesis of RNA in many plant systems (Varner & Johri, 1968). If this RNA synthesis is a primary effect of the reaction of GA_3 within a cell, it is then most likely that the primary action occurs in the nucleus. In many other plant tissues also, either an increase in protein and RNA synthesis or in the incorporation of labelled precursors into these macromolecules in the target tissues is evoked by hormones (Key, 1969).

It seems obvious that if GA interacts directly with the nucleus it would be advantageous to work with isolated nuclei or chromatin obtained from the nuclei. Therefore, attempts were made to work with nuclei isolated from light-grown dwarf peas (Varner & Johri, 1968). To determine if the RNA synthesized in the presence of GA_3 was qualitatively different, the nuclei were incubated in the presence of one of the four nucleoside triphosphates labelled in the *a* position with ^{32}P and the radioactivity then determined in the 2',3' nucleotides. It was found that a significant difference existed in the relative distribution of radioactivity among the four nucleotides of the RNA synthesized in the presence and absence of gibberellic acid. This nearest-neighbour analysis indicated that RNA synthesized by GA-treated nuclei is in part qualitatively different from the RNA synthesized by untreated nuclei.

Fractionation of the labelled RNA on MAK columns showed differences also in the types of nucleic acids produced in the presence of GA_3. The results indicated that GA preferentially enhances the synthesis of a small fraction of RNA.

However, fractionation of the labelled nucleic acids on sucrose gradients indicated that the *in vitro* synthesized RNA sediments in a peak corresponding approximately to 6–10s. These results indicate that the RNA synthesized by GA-treated nuclei has a higher average molecular weight than that synthesized by the control nuclei.

Varner and Johri (1968) concluded that the enhanced RNA synthesis could be due (a) to an increase in template DNA availability, (b) to an increase in the RNA polymerase activity, or (c) to a combination of the above two possibilities. An enhanced activity of RNA polymerase could be due to activation or to protection against degradation of the enzyme. Another explanation might be that the hormone changes the permeability of the nuclear membrane and thereby makes the RNA precursors more available for use in transcription.

Gibberellin Control of Enzyme Secretion

The gibberellin-enhanced synthesis of hydrolases is preceded by a proliferation of rough endoplasmic reticulum (Yomo & Varner, 1971).

The control of rough endoplasmic reticulum seems more likely than the control of hydrolase synthesis and secretion as a candidate for the primary site of gibberellin action since it could more readily be a general response of plant tissues to gibberellins.

Varner and his associates (Evins & Varner, 1971; Koehler & Varner, 1973) have therefore examined rough endoplasmic reticulum in detail.

There is a gibberellinenhanced incorporation of labelled choline into the lipid-soluble material of the endoplasmic reticulum and of other cell fractions. An enhanced incorporation of ^{32}P into the phosphoryl-lipids of several cell fractions is also seen. This enhancement is first measurable about four hours after the addition of gibberellin and is proportional to the logarithm of the concentration of the hormone.

GA_3 also enhances the activity of phosphorylcholine cytidylate-transferase and phosphorylcholine-glyceride transferase (Johnson & Kende, 1971).

These responses are measurable as early as two hours after treatment. Gibberellin-evoked increases in these membrane-bound enzyme activities are prevented by the inhibitors, abscisic acid, actinomycin D and cycloheximide. These early effects on phosphoryl-lipid metabolism suggest that hydrolase synthesis occurs only on polysomes bound to newly synthesized endoplasmic reticulum.

Thus the synthesis of hydrolase before the addition of GA_3 may be limited not by messenger-RNA but by the availability of appropriate membranes for the attachment of polysomes which carry hydrolase-specific messenger-RNAs.

An examination of the secretion of hydrolases is of interest because it appears that gibberellin is required for the secretion of *b* -glucanase which had been synthesized before hormone addition. Gibberellin-enhanced release of *b* -glucanase into the medium surrounding the aleurone cells is prevented by inhibitors and uncouplers of oxidative phosphorylation, by actinomycin D, 6-methylpurine, and by cycloheximide. This is in contrast to the secretion of *a* -amylase which requires phosphorylative energy transformation but is not affected by 6-methylpurine nor by cycloheximide.

These differences in the secretion of *b* -glucanase and *a* -amylase may represent different modes of secretion for the two enzymes. Or, it may be that the cellular apparatus required for the secretion of both enzymes is made during the first few hours of gibberellin treatment.

Effects of Gibberellins on Membranes

For the past few years Paleg and his associates have studied the effects of GA_3 on various membranes and membrane properties (Wood & Paleg, 1974; Wood *et al.*, 1974). Their studies were initially encouraged by the work of Bangham *et al.* (1965) who provided evidence that steroids have potent effects on model membranes composed only of lipids. Correlation between steroid activity *in vivo* and in the model membrane systems was good. A suggestion was made that steroids produced their biological effect through a direct interaction with lipid, independent of polysaccharides, proteins or even active cell metabolism.

The influence of GA_3 on the permeability of lipid model systems was first described by Wood and Paleg (1972) who suggested that gibberellins might similarly exert their physiological action through an effect on lipids. Recently, Wood *et al.* (1974) have demonstrated complex formation between GA_3 and phosphatidyl choline thus providing definitive evidence of a molecular association between the hormone and a common membrane component.

In this study, complex formation was dependent on hormone concentration. Wood and Paleg (1974) also showed that the process of glucose diffusion from liposomes (i.e. microvesicles composed of a bimolecular lipid membrane surrounding aqueous contents) is not affected by the presence of GA_3.

The same energy is required for a mole of glucose to permeate the lipid at each temperature in the presence or absence of the hormone. However the barrier to diffusion is perturbed in such a way that more glucose can pass per unit of time in the presence of GA_3. Furthermore, the temperature at which the lipid membrane undergoes a change in conformation (the transition point) is lowered by several degrees by hormone treatment.

By comparison with living systems, GA_3 could, by its influence on lipid polarity, affect two main processes *in vivo;* (a) the permeability of the membranes to regulatory ions which act as limited substrates; and (b) the activity of membrane-bound enzymes such as ATPase which are activated by sodium, potassium, etc.

Both of these effects could be brought about by the same mechanism—greater membrane fluidity.

Since GA_3 influences not only the transition point at which the liposome changes in structure but also the diffusion rate of glucose above this transition point, one or both of two mechanisms may explain its action. Either its presence directly reduces the viscosity of the central core of the membrane or, more likely, it perturbs the still-ordered polar regions of the liquid crystal phase leading to greater fluidity.

However, the temperature zone of greatest physiological interest is that where the untreated system is below the transition point while in the presence of hormone the system has switched to high temperature sensitivity.

The relationship between biological activity of different gibberellins and their ability to influence the thermal transition and membrane permeability has not been explored extensively.

However, the methyl ester of gibberellic acid, a derivative with only about 10^{-3} of the biological activity of unsubstituted GA, was found to be without effect on the temperature at which the transition of the membrane occurs. In this regard it is of interest that it also induces very little alteration of the nuclear magnetic resonance spectrum of lecithin, which indicates little complex formation between the hormone and the fatty acid.

This new area of research may suggest that the mechanism of GA action is through the direct alteration of membrane structure and/or function.

SUMMARY

The gibberellins comprise a complex and large group of related compounds which control cell elongation and enzyme secretion. The mechanism by which the hormone turns on the synthesis of enzymes has been studied in fairly close detail. Initially, it was felt that the hormone regulated the synthesis of RNA at the level of transcription.

Through the careful and dedicated work of Varner and his associates it appears reasonably clear that the hormone does not control the secretion of enzymes by controlling the synthesis of messenger-RNA.

However, it is clear primarily from the work on amylase, that inhibitors of nucleic acid synthesis and protein synthesis disrupt the production of the enzyme. The recent experiments of Paleg and his associates bring a new dimension to the question of gibberellin action by their demonstration that GA_3 affects the membrane permeability of fluidity. Thus, a new range of possible hormonal mechanisms can now be envisaged and explored.

ACTION OF CYTOKININS

The term cytokinin has been accepted practically universally as a generic name for substances which promote cell division and exert other growth regulatory functions in the same manner as kinetin. The synthesis and testing of compounds for cytokinin activity began with the discovery of kinetin (6-furfurylaminopurine). Today there are probably at least one hundred known synthetic and native cytokinins.

Structural requirements for a high order of cytokinin activity generally include an adenine molecule with the purine ring intact and with an N^6 substituent of moderate size. One exception to the requirement for a modified purine exists, notably diphenylurea and its derivatives.

Certain other substances which lack a true purine ring, such as 8-azakinetin, 6-benzylamino-8-azapurine and 6-(3-methyl-2-butenylamino)-8-azapurine are active, but each of them is less than 10% as active as its corresponding purine derivative. Substitution of either O or S for N in the N^6 position of adenine in each case results in a more than 90% loss in growth promoting activity in the tobacco bio-assay.

Cytokinins play a role in practically all phases of plant development from cell division to the formation of flowers and fruits. They affect metabolism including the activities of enzymes and the biosynthesis of growth factors.

They also influence the biogenesis of organelles and the flow of assimilates and nutrients within the plant. Cytokinins defer senescence and may protect the plant against adverse environments, such as water stress. These many diverse effects presumably stem from some primary anabolic function of the cytokinins that remains to be elucidated.

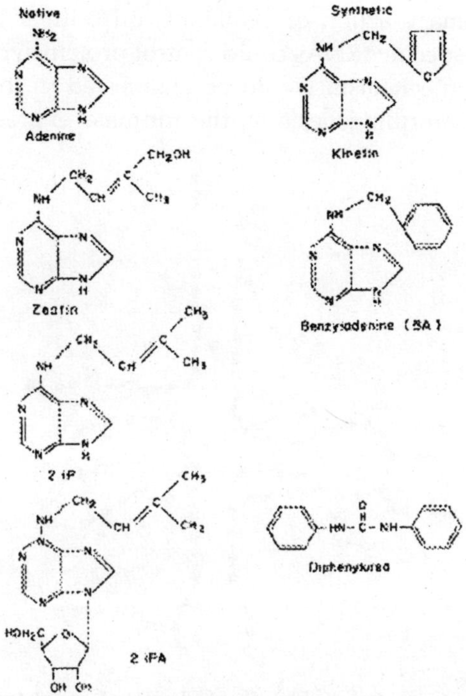

Figure: Structure of some common native and synthetic cytokinins.

Cytokinins have been found in certain transfer-RNA (t-RNA) molecules from a large number of organisms including bacteria, yeast, plants and animals (see Cherry & Anderson, 1971). In sequence analysis the native cytokinin, isopentenyl adenine, always occurs adjacent (3' side) to the anticodon. Furthermore, the location of the isopentenyl group is required for the specific tRNA to function efficiently in protein synthesis.

Results involving bacteriophage infection of bacteria indicate that cytokinin in tRNA may have an important regulatory function in protein synthesis. In soybean seedlings (Anderson & Cherry, 1969) the application of a cytokinin (6-benzyladenine) results in increased levels of two species of leucyl-tRNAs. Although the relationship of cytokinins to tRNA may not be directly associated with the primary action of cytokinins on cellular activity, the alteration of specific tRNAs could control protein synthesis. Such a regulatory mechanism would be manifested in the control of growth and morphogenesis as the ultimate expression of the hormone.

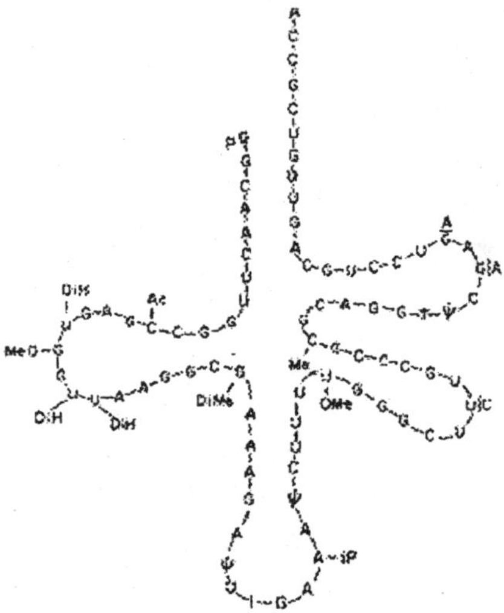

Figure: Diagram of the cloverleaf shape of seryl-transfer RNA. The IPA residue is located adjacent to the anticodon.

Actions of Cytokinins on Enzymes

Cytokinins have been shown to influence the activities of a number of specific enzymes. A most thoroughly studied case is the promotion by cytokinins of tyramine methylpherase activity in roots of germinating barley embryos. This appears to involve *de novo* synthesis although this conclusion is based solely on the results of inhibitor experiments (Mann *et al.*, 1963; Steinhart *et al.*, 1964). Activities of four other enzymes tested were unaffected by kinetin treatment.

The selective action on methylpherase is most likely related to the developmental stage of the seedling material and not to a specific affinity of cytokinin for the enzyme. Substances other than cytokinins (auxins and lysine) which also stimulated the appearance of tyramine methylpherase activity either were less effective or acted only during the early stages of germination. An analogous effect of kinetin in promoting synthesis of alkaloids in the roots of lupin has also been reported (see Skoog & Armstrong, 1970).

The induction of *a* -amylase activity in half-seeds without embryos, a well known effect of gibberellins in barley, is achieved in wheat also by treatment with cytokinins (Boothby & Wright, 1962). However, the effect of cytokinins on wheat is much less marked than that of gibberellin in the barley. In barley seeds, cytokinins cannot replace gibberellins, but do counteract an inhibition of gibberellin action by abscisic acid (Khan, 1969). The formation of amylase activity in excised hypocotyls of pea is stimulated by cytokinins but not by gibberellins or auxins (Chum, 1967). In germinating squash seeds cytokinins, but not gibberellins, at least partly substitute for the embryo in promoting the formation of isocitrate lyase and proteinase activity in the cotyledons.

The synthesis of deoxyisoflavones in soybean callus tissue is promoted by both cytokinins and auxins (Miller, 1969). The ease with which the compounds can be detected and the rapidity of the response suggests that this system might be useful as a rapid bioassay for cytokinins.

The rates of synthesis of ribulose bisphosphate carboxylase and NADP-dependent glyceraldehyde-phosphate dehydrogenase, studied in rice seedlings, appear to be under cytokinin control (Feierabend, 1969). Kinetin did not affect these enzymes until 96 hours after the beginning of germination. By contrast the activity of the cytoplasmic enzyme 6-phosphogluconic acid dehydrogenase was promoted by kinetin only during early stages of germination. It was concluded that cytokinins probably are not directly involved in gene derepression although the level of cytokinin may determine the extent to which the appropriate genes are expressed.

NUCLEIC ACID SYNTHESIS

Early studies with tobacco pith tissue showed that a proportion of the tetraploid cells present would be stimulated by kinetin treatments to enter mitosis and undergo cytokinesis without incorporation of thymidine. Thus the effect of kinetin on cell division was not always associated with DNA synthesis. In *Lemna* cultures 6-benzyl-aminopurine reverses the effect of abscisic acid in inhibiting growth and nucleic acid synthesis (van Overbeek *et al.*, 1967). DNA synthesis in response to these growth substances appears to be more rapid than RNA synthesis. In axillary buds of tobacco which are normally inhibited by the apex, treatment with 6-benzylaminopurine initiates DNA synthesis and growth (see Skoog & Armstrong, 1970).

Some inhibitors such as chloromycetin and 6-azauracil uncouple the two processes, preventing growth but not the cytokinin-induced DNA synthesis. Nevertheless, in view of the early work, it seems likely that cytokinin effects on DNA synthesis are indirect. In a detailed study of tuberization, Palmer and Smith (1969) have shown that, in excised stolons of potato, cytokinins inhibit stolon elongation and induce tuber formation. Other factors which inhibit elongation may also increase the inductive response in response to cytokinins, but they are not themselves inductive factors.

Inhibitor studies suggest that cytokinin action is associated with protein synthesis or activation but not with DNA or messenger-RNA synthesis. Effects of cytokinin on RNA and protein synthesis and isolated cell organelles have been reported, but have not yet been analyzed in depth (Skoog & Armstrong, 1970). Nuclei preparations separated from coconut milk and incubated in the presence of kinetin showed an increased incorporation of labelled precursors into RNA and protein (Datta & Sen, 1965).

The effect of kinetin on RNA synthesis in an *in vitro* system containing purified chromatin and *E. coli* RNA polymerase has been studied by Matthysse and Phillips (1969). In the presence of a particular protein fraction, kinetin stimulated RNA synthesis. A similar effect of auxin was also mediated by a protein fraction. The effects of cytokinin in this system have not yet been studied in great detail. Fellenberg (1969) has observed that *in vitro* bindnig of kinetin to chromatin of pea epicotyls caused a slight decrease in the melting point of the nucleoprotein complex.

Role of Cytokinins on Transfer RNA

One suggestion has been that the endogenous cytokinins somehow exert their activity by being incorporated into transfer-RNA (Skoog & Armstrong, 1969). For example, the synthetic cytokinin, benzyladenine, was reported to be incorporated into the tRNA of callus tissue (Fox & Chen, 1967). Richmond *et al.* (1970) specifically looked for incorporation of labelled benzyladenine into the RNA of senescing leaves, but found no evidence for incorporation. On the other hand, under similar conditions kinetin was incorporated into the RNA. Recently Walker *et al.* (1974) have shown that double-labelled benzyladenine is incorporated into tRNA at a frequency of one per 10^4 molecules.

Chen and Hall (1969) have shown that cytokinin-dependent tobacco callus tissue contains the enzyme system which attaches the D^2 -isopentenyl group to the appropriate adenine residue in tRNA. In the case of tobacco callus tissue grown in the presence of radioactive benzyladenine, the tRNA still contained its complement of isopentenyl adenine derivatives even though the benzyladenine was incorporated into the tRNA. In general, various purine analogues can be incorporated into RNA in a rather non-specific way. It is also quite possible that incorporation occurs into RNA fractions other than tRNA, and unless the tRNA is rigorously characterized one cannot exclude the possibility of error arising by contamination with other RNA fractions. To establish whether the incorporation of N^6 -substituted adenosine derivatives into tRNA has a biological significance it would be necessary to fractionate the tRNA and identify the incorporated material at a specific location in the tRNA sequence.

Model for Cytokinin Action

Cytokinins are found in transfer-RNAs of animals, microorganisms and plants (see Cherry, 1973). Furthermore, in those cases where the nucleotide sequence is known the native cytokinin, IPA (isopentenyladenosine), or a modified form, is found adjacent to the anticodon. At the moment it appears that all IPA containing species of tRNA recognize codon triplets which begin with U (Skoog & Armstrong 1970).

Thus, the anticodon loop of a cytokinin-containing RNA could be one of at least sixteen codon sequences of the following type: A— —IPA. By eliminating the nonsense terminator and the phenylalanine codons, it seems that IPA-containing tRNAs would be confined to less than ten codons.

Table. Codons beginning with U.

1st	2nd → U	C	A	G	3rd
U	PHE	SER	TYR	CYS	U
	PHE	SER	TYR	CYS	C
	LEU	SER	Ochre	Nonsense	A
	LEU	SER	Amber	TRP	G

It may be that IPA adjacent to the anticodon of these tRNA species plays some role in messenger-RNA-tRNA-ribosome recognition, and that these tRNA species are required for the translation of some messenger-RNAs into specific proteins. Based on these assumptions it is possible to speculate on the mechanism of action of cytokinins in plants.

From the observation that cytokinins appear not to be incorporated into tRNA, as precursor units, but rather the isopentenyl group comes from mevalonic acid (Chen & Hall, 1969), it seems likely that cytokinins have no effect on the synthesis of tRNA containing IPA. Furthermore, a number of chemicals, including various isomers of IPA, kinetin, 6-benzyladenine and even diphenylurea, have cytokinin biological activities, and it appears that these cytokinins do not participate in tRNA synthesis. If this is true, then what is the mechanism of cytokinin action?

The IPA would provide a unique attachment for the enzyme to bind to and subsequently to break the phosphodiester bonds. The action of this enzyme would destroy the function of the tRNA and yield free cytokinin. Cytokinins including 6-benzyladenine and diphenylurea added exogenously to plants or tissue cultures would defer senescence by essentially protecting the cytokinin-containing tRNA species. This protection would be mediated by the exogenously added cytokinin binding to the nuclease and competitively inhibiting its action.

Alternatively, the cytokinin could bind to other agents which in turn prevent an increase in nuclease activity which would destroy the cytokinin-containing tRNA. Therefore, tissue treated with cytokinin would retain its IPA-containing tRNA species and continue to synthesize essential proteins as long as sufficient free cytokinin was present within the tissue.

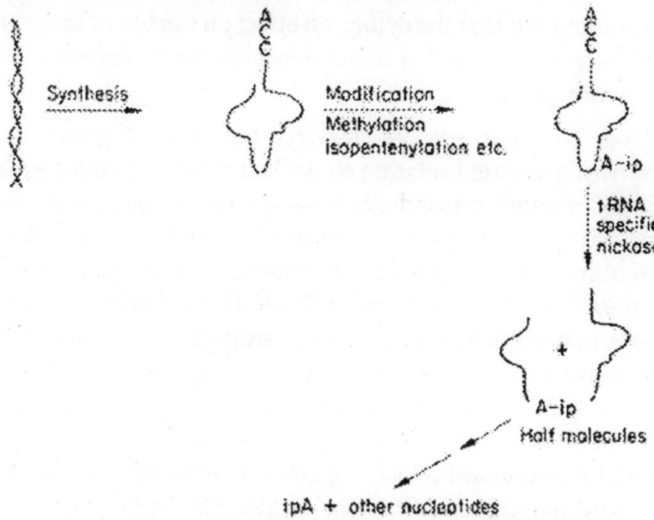

Figure: A model of cytokinin action. Based on the fact that tRNA specific nucleases are known, it is suggested that free cytokinins protect cytokinin-containing tRNAs by inhibiting nuclease activity, specific as well as general.

Preliminary unpublished work of Locy (1974) provides suggestive evidence of a tRNA-specific nuclease. We have found that chloroplasts of soybeans contain a nuclease which specifically degrades one of the three tyrosyl-tRNAs. At the moment we do not have any evidence to suggest that free cytokinin controls the activity of this tRNA nuclease. In regards to this model it is to be emphasized that it is hypothetical and is presented as a working base upon which to investigate further the mechanism of action of cytokinins at the molecular level.

SUMMARY

Cytokinins have been shown to increase the activities of a number of enzymes. These increases in activity do not appear to be a direct effect of the cytokinin but rather to be an indirect effect of some kind. Cytokinins increase the amount of DNA produced in a number of tissues. Furthermore, cytokinins increase the rate of RNA synthesis in ageing tissue. However, a number of studies have pointed out that the cytokinin effect on nucleic acid synthesis most likely is not due to a direct effect on the derepression of the DNA within the genome.

Over the past few years there has been a great deal of excitement generated relating to the fact that the native cytokinin is found in some transfer-RNA isoaccepting species. For a few years data was presented which showed that cytokinins, particularly the synthetic cytokinin, 6-benzyladenine, was incorporated directly into transfer-RNA. That excitement, however, appears to be shortlived since other studies have shown that the incorporation of the labelled synthetic cytokinins is not into tRNA species but into heavier forms of RNA. Therefore, it appears quite clear that cytokinins are likely not incorporated into tRNA directly. A model is presented which suggests that the free cytokinins act as a protective agent within the target cells of plants to protect the transfer-RNA molecules which contain the native cytokinin. It is suggested that there are specific nucleases which degrade only certain species of transfer RNA. The model implies that free cytokinin inhibits the nucleases through product inhibition, or possibly through some other mechanism which prevents the enzyme from attacking the tRNAs which contain the cytokinin. At present there is only a small amount of evidence to substantiate this model.

ACTIONS OF ETHYLENE

Ethylene is the simplest hormone which regulates plant growth. It is a natural constituent of plants and affects a wide array of physiological processes. For many years, however, investigations of ethylene physiology were concerned primarily with fruit ripening. With the advent of gas chromatography many experimenters have begun research on the biochemical and physiological action of ethylene. The production of ethylene by germinating seeds and seedlings suggests that the hormone may be involved in the normal regulation of growth and development. Evidence for such a proposal includes the fact that low concentrations of applied ethylene blocks photo-induced apical bud expansion and hook opening in etiolated pea seedlings. It was shown that apical tissues of etiolated seedlings are the major site of ethylene production. Light has been found to decrease the tendency of pea stem segments to produce ethylene in response to high concentration of auxin.

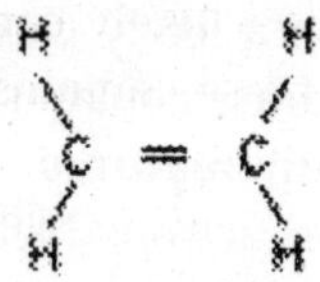

Figure: Structure of ethylene.

On the other hand, red light appears to stimulate ethylene production in dormancy. Another important role of endogenous ethylene in etiolated seedlings is the regulation of radial expansion of the pea epicotyl in a region below the apical hook. The exposure of the plant to red light or to CO_2 inhibits an ethylene-mediated increase in the diameter of the epicotyl.

Growth, flowering, abscission and fruit ripening all are affected by ethylene. It is currently popular to speculate that the mode of action of ethylene involves a mechanism which regulates some aspect of the transcription of DNA or the translation of RNA. Studies of the effects of ethylene on abscission and growth indicate sizeable changes in RNA and protein contents. There is indeed evidence of changes in the activities of catalase, peroxidase, and other hydrolases. Furthermore a study showed that exposure of soybean plants to ethylene significantly altered RNA polymerase activity associated with chromatin. As judged by nearest neighbour analysis the RNA produced from ethylene treated plants has a different base composition than that from the control. Even though ethylene at very low concentrations affects a wide array of physiological and biochemical processes in plants, much more work is required to know the mechanism of action.

Effect of Ethylene on Enzymes

A number of investigators have examined the possibility that ethylene has a direct effect on enzyme activity (Abeles, 1973). However, investigations with *b* -glucosidase, emulsin, *a* -amylase, invertase, peroxidase, and adenosine triphosphatase have shown that ethylene has no effect on these enzymes. Results reported by Abeles (1973) have similarly shown that ethylene has no effect on cellulase and carbonic anhydrase. Carbonic anhydrase was chosen because it appeared to be a likely candidate to show a positive response. Carbonic anhydrase contains zinc and has the ability to combine with CO_2, both features which suggest potential sensitivities for ethylene action. Nelson (1939) reported that ethylene increased the activity of trypsin. However, this effect was thought to be due to the removal of O_2 since H_2 had the same effect.

Actions on Membranes

Since ethylene is more soluble in oil than in water and since membranes contain large amounts of lipid material, a number of investigators have tested the idea that ethylene affects some aspect of membrane permeability. However, proponents of this idea have failed to note that CO, an ethylene analogue, does not share the same lipid solubilities as ethylene but nevertheless has similar physiological activities. The idea that ethylene has a disruptive effect on the membrane causing a change in permeability and an alteration in compartmentalization does not appear valid. Ripening fruits exhibit obvious changes in terms of permeability and retention of soluble components and it seems natural to suggest that ethylene leads to changes in membrane permeability which in turn causes softening and increased respiration. However, current evidence suggests that a change in membrane properties is an effect of ripening rather than a cause.

A similar situation exists in flowers. Nichols (1968) pointed out that solute leakage increased from carnations during senescence. Senescence and leakage were promoted by ethylene and reversed by CO_2. Ethylene has no influence on membrane permeability of potato, pea, avocado, banana, and bean. However, on the other hand, von Guttenberg and Beythien (1951) reported that ethylene increased the rate of deplasmolysis of leaves, but Burg (1968), was unable to confirm their results. Nord and Franke (1928) claimed that ethylene increased the permeability of yeast cells but the effects were small, while Abeles (1973) failed to confirm their results. Even if ethylene regulated permeability of yeast cells it would tell us little about the hormonal mechanism of action since ethylene has no effect yet detected on the physiology of yeast cells. The same criticism is applicable to the work involving red blood cells where ethylene was found to increase permeability. In this case nitrous oxide had the same effect on permeability showing that the ethylene effect was nonspecific (Lyons & Pratt, 1964). Similarly, early reports on the response of mitochondria to ethylene suggested an influence of a high concentration of the gas on conformation (Olson & Spencer, 1968). While the effects were real, subsequent work pointed out the effects were not typical of normal ethylene action. First of all, high concentrations of ethylene were required to induce conformational changes and second, saturated gases such as ethane and propane gave the same effects as ethylene (Ku & Leopold, 1970).

Enzyme Induction by Ethylene

Regeimbal and Harvey (1927) were the first to report that ethylene-treated tissues contained higher levels of particular enzymes than the control. They found that ethylene increased the level of protease and invertase extracted from pineapple fruits. Since that time, reports (see Abeles, 1973) have appeared which show effects of ethylene on a number of enzymes. This list includes acid phosphatase, ATPase, *a* -amylase, cellulase, chitinase, chlorophyllase, cinnamic acid 4-hydroxylase, cytochrome *c* reductase, diaphorase, *b* -1,3-glucanase, invertase, malic enzyme, pectin esterase, peroxidase, phenylalanine ammonia-lyase, polygalacturonase, polyphenyloxidase, protease and pyruvic carboxylase.

Enzyme induction does not always depend entirely on the action of ethylene. In some cases, cutting the tissue causes an increase in enzyme activity. The function of ethylene is to reduce the lag time or increase the rate of increase in enzyme activity. Examples of enzymes whose increased activity does not strictly depend on ethylene action include *b* -1,3-glucanase, malic enzyme, phenylalanine ammonia-lyase and peroxidase (Abeles, 1973). Excising tissue can cause wound ethylene production and it is possible that this source of ethylene plays some role in enzyme induction in tissue slices. On the other hand, enzyme induction during abscission was dependent on ethylene and for a reasonable period of time no cellulase synthesis or abscission occurred following excision unless ethylene was added to the gaseous phase.

Effects of Ethylene on RNA Synthesis

The first report of ethylene effects on RNA synthesis was by Turkova *et al.* (1965) who reported an increase in RNA synthesis during epinasty of tomato leaves. Whether or not RNA synthesis was required for epinasty was not shown, although the idea is intriguing. Studies with actinomycin D indicated that RNA synthesis occurred during abscission and was required for the process to occur (Abeles & Holm, 1966). Support for this interpretation stems from the work of a number of investigators (see Abeles, 1973). It is known that the increase in RNA synthesis precedes that in protein synthesis and is localized at or near the separation layer. The increase in RNA occurred in all fractions including messenger-RNA, ribosomal-RNA and transfer-RNA. Differential extraction of the nucleic acids indicated that the ethylene-stimulated fraction was confined to that portion of RNA extracted by sodium lauryl sulphate with the increase found in ribosomal- and messenger-RNAs.

The inhibitor, 5-fluorouracil, which blocks 50% of the ethylene-enhanced incorporation of ^{32}P into RNA did not inhibit abscission.

The greatest inhibition occurred in transfer-RNA and ribosomal-RNA which indicates that the synthesis of all fractions of RNA is not required for abscission. Presumably as long as messenger-RNA was being synthesized, enough ribosomal- and transfer-RNAs were already available within the cell to permit abscission to take place. When all RNA synthesis was blocked by actinomycin D, abscission stopped (Holm & Abeles, 1967). Ethylene has also been found to promote RNA synthesis in preclimacteric fruit. Marei and Romani (1971) reported, as in the case of abscission, that the synthesis of all classes of RNA in fig fruit was stimulated. Holm *et al.* (1971) found that ethylene increased RNA synthesis in apples and that the increase in RNA was followed by an increase in protein synthesis. However, Sacher and Salminen (1969) reported that they failed to find an increase in RNA synthesis when preclimacteric bananas or avocados were treated with ethylene.

Effects of Ethylene on Chromatin Activity

Holm *et al.* (1970) have reported that ethylene inhibited the growth of the apical part of soybean seedlings and caused an increase in the elongating and basal portions of the stem. At the same time, RNA levels in the apical zone were reduced while the levels were enhanced in the elongating and basal regions. Chromatin from the various parts of the seedling were studied to determine if the capacity for RNA synthesis was also modified. They found that the activity in the apex was reduced while the activity in the elongating and basal regions were promoted. The rate of the response was rapid since the increase in chromatin activity was apparent after 3 hours. Nearest-neighbour analysis of the RNA synthesized demonstrated that there was a qualitative difference between RNA synthesized from chromatins of normal and ethylene-treated tissue. It was concluded that ethylene can regulate RNA synthesis as manifested by a change in quantity and kind of RNA.

Effects of Ethylene on DNA Metabolism

Plant growth is either promoted, inhibited or unaffected by ethylene depending upon the tissue involved. Examples of growth promotion are swelling, epinasty, hook closure, seedling elongation and seed germination (Abeles, 1973). Bud break is probably a special case. Here no growth takes place as long as ethylene is present. However, after the gas is removed growth of the buds ensues. Growth inhibition is seen as arrested development of buds, leaves or apical meristems.

Mature tissue, such as stems and leaves. do not undergo any change of size or weight, although premature senescence usually occurs. Since growth, or the lack of it, may be associated with cytokinesis, it is of interest to learn that DNA synthesis is controlled by ethylene. Holm and Abeles (1967) reported that DNA synthesis or DNA content in bean leaf tissue was not affected by seven hours exposure to ethylene, although abscission was promoted.

Later they found that in soybean seedlings treated with ethylene the synthesis of DNA was inhibited in the apex where growth was inhibited but was promoted in the subapical part where swelling took place. Burg *et al.* (1971) found a similar situation in pea seedlings in which inhibition of cell division, measured as a reduction in metaphase figures, occurred within two hours after ethylene was added. However, it is not clear whether the change in DNA synthesis was the cause or the result of inhibited growth. Ethylene slows the growth of pea seedlings very quickly; for example, Warner and Leopold (1971) found growth slowed six minutes after ethylene was introduced into the gas phase and returned to normal sixteen minutes after the ethylene was removed.

Kinetic studies on changes in DNA or other postulated sites of action are thus required to establish the relationship between cause and effect. Burg *et al.* (1971) have suggested that ethylene regulates DNA synthesis by some action on microtubule structure essential for spindle fibre formation during mitosis. If the action of ethylene was directed toward microtubules, this might also explain the reorientation of cellulose microfibril deposition that occurs in swelling.

Even though little is known about the binding site(s) of ethylene, in a few cases such as ripening, abscission, swelling and senescence there is reason to believe that the combination of ethylene with a site results in the regulation of protein or enzyme synthesis which in turn accounts for the observed responses. In these cases RNA synthesis, presumably messenger-RNA, with the accompanying support of soluble- and ribosomal-RNA, is an essential or early step suggesting regulation of gene action. However, in other cases, especially those associated with the physiology of excised tissue, control is probably not exerted at the level of RNA or protein synthesis. In the case of the inhibition of elongation, the action is apparently directed toward blockage of DNA metabolism.

The site of action of ethylene in epinasty, root initiation, intumescence formation and floral initiation is even more poorly understood. The only valid conclusion appears to be that a number of essential features of plant growth and development are susceptible to ethylene action. In the final analysis it is concluded that there may be as many mechanisms of ethylene action as there are modes of ethylene operation.

ACTIONS OF ABSCISIC ACID

Abscisic acid (ABA) is a plant hormone which now ranks in importanance with the auxins, gibberellins, cytokinins, and ethylene. Interest in the physiology and chemistry of ABA has grown greatly since the structure was established in 1965. During the 1950's and early 1960's a number of laboratories were engaged in research on growth-inhibiting substances.

ABA was first isolated from cotton plants and was named abscisin II by a team from Addicott's laboratory (Ohkuma *et al.*, 1963). During the same year Wareing's research team (Eagles & Wareing, 1963) isolated an active substance from *Acer* leaves which they named dormin. Abscisin II and dormin are the same substance which is now called abscisic acid. Plant tissues of all ages appear to synthesize and inactivate ABA.

The number of various plant responses affected by ABA is very large. Generally, the physiological processes are related to senescence or abscission and growth retardation or inhibition.

ABA appears to act as an abscission accelerating hormone in many fruits and leaves. Furthermore, it also tends to induce dormancy in some woody plants.

ABA has been shown to move from the leaves to the apical bud to bring about a dormant condition. In potato, the levels of inhibitors, including ABA, decrease during the quiescent period prior to renewed growth. ABA in extremely low concentration, moreover, prolongs the dormancy of excised potato buds.

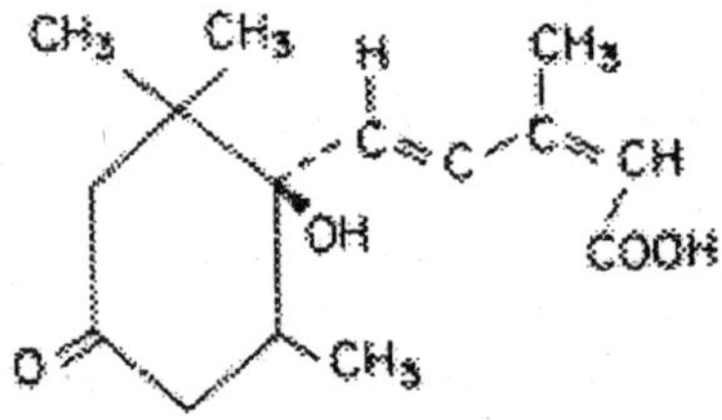

Figure: Structure of abscisic acid.

Recently, ABA has been shown to be involved in the responses of many plants to stress conditions. As the ABA concentration increases in the leaves the stomata close. In this way, it appears that ABA is directly involved in the opening and closing of the stomata, thus regulating the rate of transpiration. Through this mechanism, ABA appears to protect the plant through conditions of water stress or drought.

At present the mechanism of action of ABA is not clearly understood. However, the available evidence indicates that ABA affects transcription as shown by reduced activity of chromatin-associated RNA polymerases. In other cases the mechanism of action of ABA appears to involve regulation of the translation of long-lived messenger-RNA, whilst its effects on stomata probably involve the regulation of membrane permeability.

Role of ABA in Dormancy

One of the lines of research which led to the isolation of ABA from sycamore leaves was a change in the growth inhibitory activities in extracts of tree seedlings grown on long- or short-days. An increase in content of the growth inhibitory material in leaves was noted during the late summer and early autumn (Phillips & Wareing, 1958). Furthermore, the idea that bud dormancy in potato tubers may be caused by growth inhibiting substances had been suggested by Hemberg (1949).

The original correlation of inhibitory material in leaves grown under short- and long-days and the induction of dormancy were based on measurements obtained by bioassay techniques. Recently, Lenton *et al.* (1973) have attempted to repeat these determinations using gas chromatography to compare the amounts of ABA present.

They found that transferring birch, red maple or sycamore plants to short-days had no effect on ABA content. The importance of a balance between growth promoters and growth inhibitors has been stressed frequently. Tinklin and Schwabe (1970) have found considerably more inhibitor in bud scales than in the bud axis of blackcurrant. In these experiments the ABA concentrations were correlated with the degree of dormancy and the levels were reduced by treatments which encouraged bud break.

The most convincing evidence suggesting that ABA induces dormancy is the production of turions in *Lemna polyrrhiza.* When ABA is added to the medium under conditions that allow continuous growth, it not only inhibits growth but causes production of the dense, dormant, fronds known as turions (Stewart, 1969).

ABA is a potent inhibitor of seed germination, and its presence as a major growth inhibitor in dormant seeds of many species has cast it in the role of the maintainer of seed dormancy.

ABA has been isolated from seeds of many genera of higher plants, and seeds of an equally large number of species have been prevented from germination by soaking in ABA solutions. When seeds and fruit parts are separated, it is usually found that the concentration of ABA in the fruit is about 5 to 10 times greater than in the seed (Milborrow, 1974).

Lipe and Crane (1966) found that ABA in peach seeds decreased during stratification. Obviously, the part played by endogenous growth promoting compounds and the balance between them and ABA during the breakage of dormancy needs to be explored in more detail.

Certainly, the balance between the two kinds of regulators is important. ABA is not an irreversible inhibitor since one of its most striking features is the facility with which it can be leached out of treated seeds thereby allowing the resumption of growth.

Effects on Abscission

In the last few years several hundred experiments have been carried out in which plants, and parts of plants, of many genera have been treated in a variety of ways with ABA.

It is quite unusual for any report to show clearly that the hormone controls abscission of leaves, which leads to the conclusion that exogenously applied ABA has little effect on leaf abscission. Nevertheless, ABA was first isolated by following its abscission-accelerating activity in petiolar stumps of cotton explants.

The growth inhibitory action of this factor was believed to be responsible for the premature abscission of immature young lupin fruits (Cornforth *et al.*, 1966). Consequently, the abscission-accelerating effect of ABA is well documented and extensively discussed (Milborrow, 1974).

Many of the experiments reporting leaf abscission have been carried out on tree crops using extremely high concentrations of ABA and often near the end of the growth season in attempts to cause abscission of fruits. The observed stimulation of leaf abscission may be an indirect effect of non-physiological high concentrations which stimulate ethylene production (Edgerton, 1971).

For example, Cooper and Henry (1968) treated orange trees with sprays of 500 μg ml–L of ABA in summer and winter. The summer treatment caused leaves to develop colour and fall, but the winter treatments had no such effects. Olive trees were found to suffer some leaf abscission in one experiment but no effect was observed in another.

It appears, therefore, that ABA is not closely involved in the regulation of leaf abscission. The test system in which it has a stimulatory action consists of isolated petiolar explants containing presumptive abscission zones and maturing leaves nearing the end of their life. Even this tissue requires application of abnormally high concentrations of ABA to manifest an effect.

The role of ABA on fruit abscission is more certain. The early work of Van Steveninck (1959) showed that the inhibitor now known as ABA was intimately involved in the abortion of the young immature fruit of lupin. Application of ABA to mature peach, olive, apple and citrus fruits accelerated abscission and the effect of ABA was also marked on young grape flowers and berries (see Milborrow, 1966).

Effects of ABA on Wilting

Wright (1969) found that when cut shoots were wilted there was an increase in the concentration of the so-called *b* -inhibitor. He went on to identify this inhibitor as ABA and defined the conditions under which the increase occurred. A water loss of about 10% in total fresh weight caused approximately 40-fold increase in ABA content while further water loss had no additional effect.

Wright and Hiron (1969) have reported that other stress conditions, such as waterlogging, caused a similar rise in ABA, but they point out that such treatments cause wilting by reducing the efficiency of water uptake.

The surprising feature of the increase in ABA is the rapidity with which it occurs. The content in turgid bean leaves has been calculated from the results of Wright and Hiron (1969). They show that ABA content increases from a normal level of 6 μg kg–1 fresh weight to 7 μg kg^{-1} within 7 minutes after blowing a dry and warm air stream across the plants.

This level increases to 33 μg kg^{-1} within 25 minutes and 68 μg kg^{-1} within 45 minutes. In other experiments they show that in wheat leaves the ABA content increased from 23 to 171 μg kg^{-1} fresh weight within 4 hours of wilting. The ABA content of bean leaves remained at 67 μg kg^{-1} while they were kept waterlogged for 5 days.

The implications of these data have yet to be explored in detail, but the observations offers a feasible explanation of the reduced growth of crops suffered during drought. The extra ABA is probably formed by synthesis rather than by release from a precursor or conjugate because much more labelled mevalonate was incorporated into ABA by a sample of leaves that had been fed and then wilted than similar leaves which were kept moist during the entire experiment.

Furthermore, the presence of 40 times the amount of a precursor or conjugate probably would have been detected by extraction and bioassay techniques. Many types of experiments have shown dramatic increases in ABA content on wilting in french beans, brussel sprouts, sugar cane, wheat, avocado spinach, cotton, peas and tomatoes (see Milborrow, 1974).

Perhaps the best information supporting a direct involvement in the direct closure of the stomata is afforded by a wilty tomato mutant produced by X-irradiation (Imber & Tal, 1970). The shoots of this plant contain one-tenth the amount of ABA contained in the normal variety. Applications of 0.1 to 10 μg of ABA caused a rapid and progressive reduction in transpiration rate of leaves and leaf discs. With 10 μg ABA ml^{-1} the stomata closed in darkness.

Affects of ABA on Enzyme Activities

The ability of plant hormones to affect directly the activity of enzymes has not been extensively investigated. When an effect has been reported it is smaller than would be expected if the hormone specifically regulates at that site. Hemberg (1967) reported that ABA can inhibit *a* -amylase *in vitro,* but as the enzyme used was extracted from fungi the significance as related to higher plants requires additional investigation.

ABA has been shown to complex with fungal *a* -amylase and thereby change its physical properties. Saunders and Poulson (1968) found a slight stimulation of invertase activity at 10^{-7} M

ABA and slight inhibition at 5×10^{-7} M and above. Again the significance is difficult to assess since a fungal enzyme was used.

Effects of ABA on Nucleic Acid Synthesis

The first observations of the effects of ABA on nucleic acids were made by Van Overbeek *et al.*, (1967) using *Lemna.* The incorporation of radioactive phosphate (^{32} P) into nucleic acid fractions was inhibited by ABA but reversed by benzyladenine.

The primary site of action of ABA cannot be deduced with certainty because of the time scale and the complexity of the responses. Subsequent work has indicated that DNA synthesis is almost certainly not the primary target of ABA (Villiers, 1968). This has been clearly shown to be the case in dry wheat embryos by Chen and Osborne (1970) who found that protein synthesis commenced from imbibition and was inhibited at 6 hours by ABA, whereas RNA synthesis, as measured by the incorporation of ^{3} H-uridine, was not measurable until 12 hours. In the same experiments the incorporation of ^{3} H-thymidine into DNA was measurable after 24 hours.

Experiments by Pearson and Wareing (1969) have shown that chromatin-directed RNA polymerase activity was slightly inhibited by 0.26 *m* g ml^{-1} of ABA when added to the grinding medium. However, when ABA was added to the purified chromatin little or no effect was noted. Bex (1972) also reports that ABA has no effect on the binding between nucleohistones and DNA as measured by their melting point measurements. Schwartz (1971) has shown that ABA can alter the balance of alcohol dehydrogenase in maize, but the inhibitor was added during the growth of the cells. Thus, while it appears that ABA has small and random effects on nucleic acid biosynthesis in a number of plants, it is reasonably clear that the hormone has little or no direct effect on the synthesis of nucleic acids at the level of the genome.

The Involvement of Abscisic Acid in Messenger-RNA Translation

Ihle and Dure (1972) investigated the appearance of various enzyme activities during the germination of cotton seeds and embryos. They showed that protease and iso-citrate lyase appear to be synthesized *de novo* during germination utilizing messenger-RNA which had been transcribed much earlier when only about 60% completion of embryogenesis had been reached. The translation of these messenger-RNAs during the last 40% of embryogenesis was apparently prohibited by the presence of ABA diffusing into the embryo from the ovule wall.

The mode of action of ABA in maintaining translation inhibition appears to involve RNA synthesis as judged by the fact that the ABA inhibition was inhibited by the presence of actinomycin D. Translation of the required messenger-RNAs for the germination enzymes may be induced prematurely by simply dissecting the ovules from the plant, suggesting that the breakage of the connection between the ovule and the mother plant may be required for the *in vivo* induction of translation. This hypothesis is further substantiated by the observation that, *in vivo*, the connection between the ovule and the placenta is normally severed at 60% completion of embryo-genesis.

The results coming from Dure's laboratory are the only ones at present which show that ABA may have a direct effect on the translation of messenger-RNA pre-existing in the embryo tissue.

SUMMARY

Even though ABA was first found in cotton plants and dormant sycamore leaves, it seems reasonably clear that this hormone has little or no effect on either the abscission of leaves or the dormancy of buds. It appears that the most likely physiological action of ABA on plants is to control the abscission zone in the fruit pedicel and thereby allow abscission to take place and cause fruit drop.

Secondly, ABA very definitely appears to be involved in stomatal closure in response to stress conditions, particularly drought. In terms of a more biochemical approach it appears that ABA has little effect on various enzyme activities. Those enzymes studied show only small enhancement, or in some cases, inhibition, by ABA. Generally it has been shown that ABA inhibits the synthesis of various nucleic acids.

Again, one must be drawn to the conclusion that ÁBA has little effect directly on nucleic acid synthesis at the genome level. The most excitingp aproach to abscisic acid action appears to be that found in the cotton embryo where it has been shown to inhibit the translation of pre-existing messenger-RNAs found in the developing embryos.

Even this response probably involves more complex processes than a direct inhibitory effect, since actinomycin D inhibits the ABA-induced inhibition. These data possibly imply that a suppressor molecule has to be formed to bring about effects.

4

THE PLANT PHYTOCHROME ACTION

INTRODUCTION

Light, in addition to providing the energy which drives photosynthesis, also acts as a regulatory environmental stimulus. Many light-controlled plant responses are now believed to be mediated by the photoreceptor, phytochrome. This pigment has a regulatory role in all phases of plant growth and development (photomorphogenesis) and is apparently ubiquitous in all taxonomic groups of eukaryotic plants with the exception of fungi.

The central question of the molecular mechanism of phytochrome action is, however, yet to be resolved. Approaches to this problem fall into two broad categories: (a) Examination of the physical and chemical properties, localization and behaviour of the phytochrome molecule itself; (b) Investigations of the diverse array of phytochrome-mediated plant responses demonstrable at almost any level from the molecular to the gross morphogenetic. The principal findings and concepts that have emerged from such studies are summarized next. Some of the evidence that has led to these concepts is examined in subsequent sections.

PHYTOCHROME DOGMA

Phytochrome is a blue-green biliprotein. Two principal forms of the molecule are readily distinguishable on the basis of their absorption spectra.

and biological activity. The *Pr* form (= red-absorbing, l_{Max} = 660 nm) is considered biologically inactive, whereas the *Pfr* form (= far-red-absorbing, l_{Max} = 730 nm) is considered to be the active species i.e. capable of inducing a biological response.

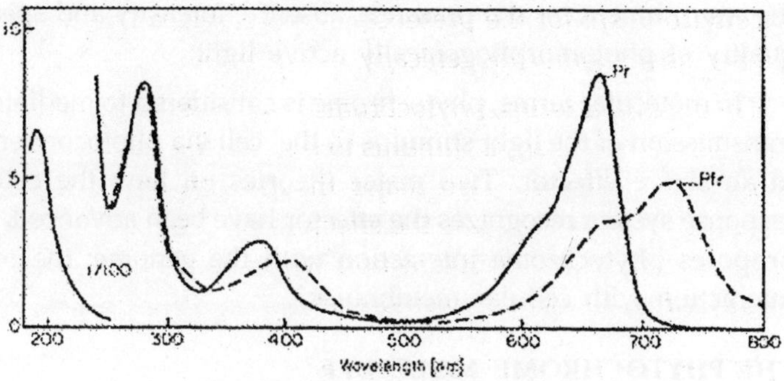

Figure: Absorption spectra of purified oat phytochrome following saturating irradiations with red and far-red light (after Anderson et al., 1970).

The two forms are reversibly interconvertible by red and far-red light ('phototransformation'). In addition, whereas the *Pr* form is stable in the dark, *Pfr* can revert thermally to *Pr* ('dark reversion') or undergo an irreversible loss of photoactivity ('destruction'). These properties are schematically represented thus:

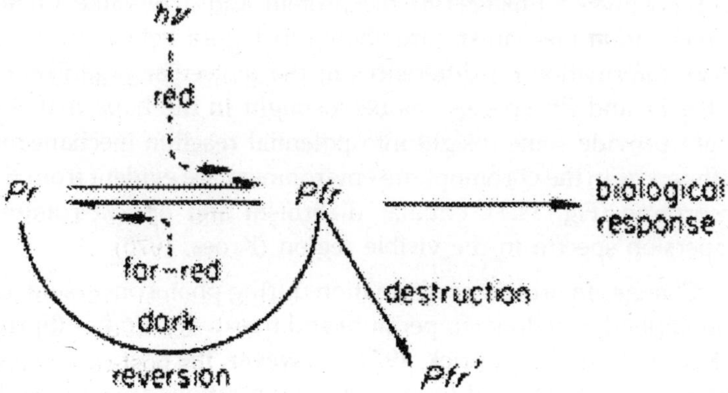

Red light absorbed by *Pr* converts the molecule to the *Pfr* form which in turn induces a measurable biological response. Far-red light, dark reversion and destruction provide alternate pathways for the removal of *Pfr* and thereby the potential for reversing induced responses. These properties endow the molecule with the capacity to function as a reversible biological switch, monitoring the environment for the presence, absence, intensity and spectral quality of photomorphogenically active light.

In molecular terms, phytochrome is considered to mediate the transmission of the light stimulus to the' cell via photoconversion to an active effector. Two major theories on how the cellular response system recognizes the effector have been advanced. One proposes phytochrome interaction with the genome; the other, interaction with cellular membranes.

THE PHYTOCHROME MOLECULE

Molecular Properties

The purified phytochrome molecule is a water-soluble chromoprotein containing less than 4% carbohydrate (Briggs & Rice, 1972). The native monomer is a polypeptide of 120,000 daltons but can form higher molecular weight aggregates. The chromophore is thought to be a linear tetrapyrrole with some evidence suggesting that there is one chromophore per monomer (Tobin & Briggs, 1973). The proposed chromophore structures, their postulated linkages to the protein and a possible photo-isomerization mechanism are shown in Figure below (Rüdiger, 1972). Information on differences in the molecular organization of the *Pr* and *Pfr* species has been sought in the hope that this might provide some insight into potential reaction mechanisms. Differences in the chromophore environment are evident from the absorption (Fig. 14.1), circular dichroism and optical rotatory dispersion spectra in the visible region (Kroes, 1970).

Changes in protein conformation during photoconversion are also implied from low temperature and freeze-dry studies (Spruit & Kendrick, 1973; Kendrick, 1974). However, the post-conversion differences in the *Pr* and *Pfr* protein configurations are apparently only quite small as revealed by a variety of spectral and chemical methods (Briggs & Rice, 1972).

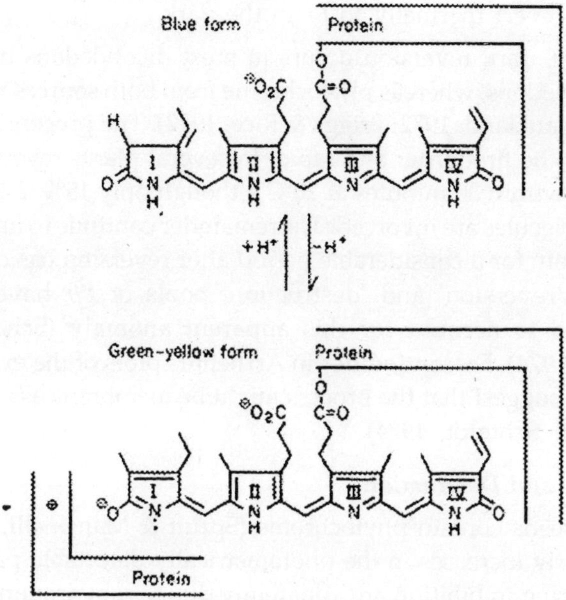

Figure: Proposed structure for the phytochrome chromophore, its linkage to the protein and possible phototransformation mechanism (after Rüdiger, 1972). The 'blue' form is thought to correspond to P **r** ***and the 'green-yellow' form to Pfr.***

Differences in surface residues are suggested by the ultraviolet difference spectrum (Tobin & Briggs, 1973); the differential reactivity of *Pr* and *Pfr* toward glutaraldehyde (Roux, 1972), and *N* -ethy maleimide (Gardner *et al.*, 1974); and the differential electrostatic binding of *Pr* and *Pfr* to ribosomal material in plant extracts (Quail, 1975b).

Dark Reactions

'Dark reversion' and 'destruction' are the so-called dark reactions of phytochrome. Neither the molecular bases nor the physiological significance of these processes is well understood.

Dark Reversion

Pr is thermodynamically stable and can only be converted to *Pfr* by light (Lhoste, 1972). *Pfr*, in contrast, is metastable and can therefore revert thermally to *Pr* in the dark.

In vivo, dark reversion occurs in most dicotyledons but not monocotyledons, whereas phytochrome from both sources reverts *in vitro* (Frankland, 1972; Briggs & Rice, 1972). The process *in vivo* appears to be first order and rapid. In several plants reversion is complete within 30 minutes at 20°C although only 15% to 20% of the *Pfr* molecules are involved. The remainder continue to undergo 'destruction' for a considerable period after reversion has ceased. Separate 'reversion' and 'destruction' pools of *Pfr* have been postulated to account for this apparent anomaly (Schäfer & Schmidt, 1974). Discontinuities in Arrhenius plots of the extent of reversion suggest that the process might be membrane associated (Schäfer & Schmidt, 1974).

Synthesis and Destruction

Dry seeds contain phytochrome (Spruit & Mancinelli, 1969). Rapid, early increases in the photometrically detectable pigment (*Ptot*) during imbibition are apparently due to rehydration of the molecule rather than synthesis (Tobin *et al.,* 1973). The pigment can be stored as either *Pr* or *Pfr* and rehydrated in the stored form (Vidaver & Hsiao, 1972). This suggests an explanation for the appearance of *Pfr* in the dark, sometimes observed in seeds (Rollin, 1972). Further increases in *Ptot* during the growth of etiolated seedlings result from *de novo* synthesis of new molecules in the *Pr* form (Quail *et al.,* 1973b). The pigment accumulates to high levels in the dark ultimately reaching a plateau.

'Destruction' is the disappearance of photometrically detectable *Pfr* without the concomitant appearance of equimolar quantities of *Pr* (Frankland, 1972). This decrease in photoactivity is paralleled by a loss of immunologically detectable phytochrome (Coleman & Pratt, 1974). Since no recycling of the protein moeity occurs, 'destruction' would appear to be a genuine degradative process (Quail *et al.,* 1973b). Some evidence suggests that this process may be enzymatic (Kidd & Pratt, 1973). Destruction is temperature dependent, but the absence of discontinuities in Arrhenius plots suggests that in contrast to dark reversion, it is not membrane associated (Schäfer & Schmidt, 1974). Destruction is observed in both monocotyledons and dicotyledons with half-times ranging from 20 minutes to 4 hours (Frankland, 1972; Schäfer *et al.*, 1973; Kidd & Pratt, 1973).

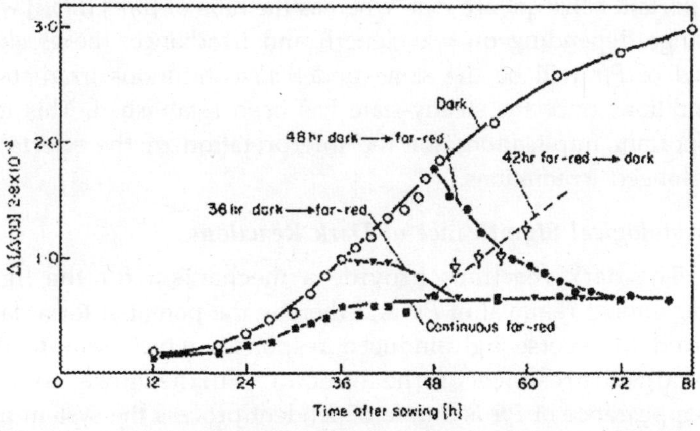

Figure 14.5: Total phytochrome levels in Sinapis cotyledons as function of time in the dark(o); continuous far-red (■); 42 h far-red ® dark (D); 36 h dark ® far-red (▲); 48 h dark ® far-red (●) (after Schäfer et al., 1972).

Destruction of *Pfr* occurs both in the dark following brief irradiations and in continuous light. In the dark, destruction rapidly removes all unreverted *Pfr* leading to a short-term decline in *Ptot*. The reduced pigment levels are replenished, however, as the dark period proceeds by *de novo* synthesis of new *Pr* molecules (Quail *et al.*, 1973b). In continuous light, initially high *Ptot* levels decline at a rate proportional to the photosteady-state *Pfr* concentration (Frankland, 1972) ultimately reaching a new plateau. This new plateau represents a steady-state equilibrium between synthesis and degradation (Schäfer *et al.*, 1972; Quail *et al.*, 1973b). *Pr* synthesis appears to be a continuous zero order process, itself unaffected by light. The *Ptot* level is then regulated against this background by the disparate first order degradation rate constants for *Pr* and *Pfr*.

The plateau level of *Ptot* established at the steady-state by prolonged irradiations is a function of both wavelength and irradiance (Schäfer & Mohr, 1974; Schäfer, 1975). The *Pfr* level, in direct contrast however, is independent of both irradiance and wavelength under these conditions. This has the extremely important consequence that, whereas the *ratio* of [*Pfr*]:[*Ptot*] will change depending on wavelength and irradiance, the *absolute* level of *Pfr* will be the same under *all* continuous irradiation conditions once the steady-state has been established. This has important implications for the interpretation of the effects of prolonged irradiations.

Physiological Significance of Dark Reactions

The dark reactions provide a mechanism for the light independent removal of *Pfr* and thereby the potential for a dark period to reverse light-induced responses which require the sustained presence of the effector. Furthermore, as the disappearance of *Pfr* is a time dependent process the system has the potential for measuring time. In principle, therefore, phytochrome should enable the plant to distinguish between light and dark and to time the dark period. Initially it was thought that this so-called 'hour-glass' principle was the basis for phytochrome-controlled photoperiodism (see Vince, 1972). Supporting evidence is scant, however, and this theory has now fallen into disrepute. Some systems do, nevertheless, respond to the timed disappearance of *Pfr* in the dark. An example is accumulation of the enzyme lipoxygenase (Oelse-Karrow & Mohr, 1973).

During prolonged irradiations, synthesis and destruction appear to have the additional role of maintaining a constant *absolute* level of *Pfr* irrespective of wavelength (Schäfer, 1975). This implies that *Pfr per se* is not the effector of high irradiance responses.

Localization

The distribution of phytochrome is highly specific at the tissue and cellular levels as determined photometrically (Briggs & Siegelman, 1965) and immunocytochemically (Pratt & Coleman, 1971; 1974). In dicotyledons, the highest pigment levels are in the apical regions. In etiolated grass shoots, large quantities occur in parenchyma cells near the tip of the coleoptile and in the rapidly differentiating tissue near the shoot apex. High levels of the pigment are also found in root cap cells.

Attempts to establish the subcellular localization of phytochrome fall into two categories:

(a) measurement of phytochrome-induced responses having a spatial or vectorial component from which the photoreceptor location can be inferred;

(b) direct measurements of the pigment itself, either *in situ* or in subcellular fractions.

The pattern of chloroplast movement in the alga *Mougeotia* in response to polarized red and far-red microbeams has been taken as evidence that phytochrome located and orienled on or near the plasmalemma controls this response (Haupt, 1972b). The directional growth of the germ tubes of *Dryopteris* in polarized light has been interpreted similarly (Etzold, 1965). The change in ion flux associated with phytochrome mediated leaflet movement (Satter & Galston, 1973), root tip adhesion to glass surfaces (Tanada, 1968), and changes in bioelectric potentials (Newman & Briggs, 1972) are also indicative of phytochromecontrolled changes in surface properties but not necessarily that the pigment is a permanent membrane component.

A report of phytochrome-controlled development of isolated etioplasts *in vitro* (Wellburn & Wellburn, 1973) implies the presence of functionally active pigment in or on the organelles. More recently a rapid, phytochrome-mediated change in the level of gibberellin extractable from an etioplast-rich fraction in response to *in vitro* irradiations has been demonstrated (Evans, 1975; Evans & Smith, 1976a; Cooke & Saunders, 1975). A red/far-red reversible reduction of NADP *in vitro* in response to irradiation of a mitochondria-rich fraction has also been reported (Manabe & Furuya, 1974). Furthermore, phytochrome has been detected spectrophotometrically in both etioplast- (Evans & Smith, 1975) and mitochondria-rich fractions (Manabe & Furuya, 1974).

Both spectrophotometric and immunological techniques have been used for direct, *in situ* measurements of the intracellular distribution of phytochrome. The presence of the pigment in the nucleus has been claimed on the basis of microspectrophotometric scans of cells (Galston, 1968) but these data have been challenged (Kendrick & Spruit, 1972; Tobin *et al.*, 1973). The cytochemical visualization of phytochrome antibody in non-irradiated tissue sections indicates a general distribution of the photoreceptor throughout the cytoplasm in addition to an association with nuclei and plastids (Pratt & Coleman, 1971). Brief red irradiation prior to fixation causes the pigment in some tissues to concentrate in discrete, as yet unidentified regions of the cytoplasm (Mackenzie *et al.*, 1974).

Non-saturating irradiations of maize coleoptile segments with red and far-red light polarized normal to the longitudinal axis were found to photoconvert about 20% more phytochrome than when polarized parallel to this axis (Marmé & Schäfer, 1972). This was interpreted as indicating that phytochrome is located and oriented in the plasmalemma. In considering the locational and orientational rigidity of phytochrome implied from polarized light studies the known rapid and highly fluid lateral and rotational diffusion of other membrane proteins should be borne in mind (Cone, 1972; Singer, 1974).

Cell fractionation procedures have also been used in attempts to localize phytochrome in subcellular components. The well-known precipitation of the pigment protein at low pHs (6.2) (Siegelman & Butler, 1965; Hillman, 1967) has been overlooked in some studies leading to claims of associations of phytochrome with mitochondria (Gordon, 1961) and plasmalemma (Marmé *et al.*, 1971).

Little of the pigment (< 10%) sediments from homogenates of nonirradiated tissue at neutral pH at forces up to 144,000 × g (Rubinstein *et al.*, 1969; Siegelman & Butler, 1965). Red irradiation prior to extraction, however, substantially enhances the level of phytochrome subsequently associated with pelletable material (Quail *et al.*, 1973a).

Irradiation of extracts from dark grown material has a similar effect (Marmé *et al.*, 1973). Initially claims were made of the isolation of a phytochrome-containing membrane fraction that could be reversibly 'solubilized' by withdrawal of Mg^{2+} (Marmé *et al.*, 1973; 1974).

More recently, however, the pigment in this fraction has been shown to be associated with degraded ribonucleoprotein (RNP) material, probably of ribosomal origin (Quail, 1975b). This association apparently results from the preferential electrostatic adsorption of *Pfr* onto ribosomal material—either free or membrane-bound in the endoplasmic reticulum (Williamson *et al.*, 1975). Whether such an association is artefactual or biologically meaningful is yet to be established. A recent promising variation on this approach is the use of glutaraldehyde in an attempt to immobilize the pigment in the cell prior to extraction (Yu, 1975.)

The existence of meaningful phytochrome-membrane interactions are by no means excluded by the above findings. It has been shown, for example, that phytochrome can mediate photoreversible conductance changes in artificial lipid membranes (Roux & Yguerabide, 1973). The suggestion that phytochrome might function as a stereospecific protein ligand capable of interaction with cellular membranes has been made (Quail & Schäfer, 1974; Boisard *et al.*, 1974).

PHYTOCHROME PHYSIOLOGY

The biological responses attributed to phytochrome can be usefully characterized in terms of three different but interrelated sets of criteria:

(a) the nature of the involvement of light in the inductive process;

(b) the temporal expression of the induced response; and

(c) the type of cellular or developmental process affected.

Induction-Reversion and High Irradiance Responses

Two types of light-controlled phenomena have been attributed to phytochrome—the so-called 'induction-reversion' and 'high irradiance' (HIR) responses (Mohr, 1972). This terminology arises from the irradiation conditions under which the responses are observed and suggests a fundamental difference in the manner in which the phytochrome molecule transmits the light signal to the cell in each case. It does not necessarily reflect an intrinsic property of the actual, biological parameter being monitored. Some parameters display both modes of response, others only one.

Induction-Reversion Responses

These are the classical phytochrome responses (Borthwick, 1972). A change in the biological parameter being monitored is induced by a brief irradiation of low intensity red light and reversed by a subsequent far-red pulse.

The accumulation of anthocyanin in *Sinapis* illustrates this point. Another well known example is lettuce seed germination. This is repeatedly photoreversible for up to 100 alternate red and far-red irradiations (Borthwick, 1972).

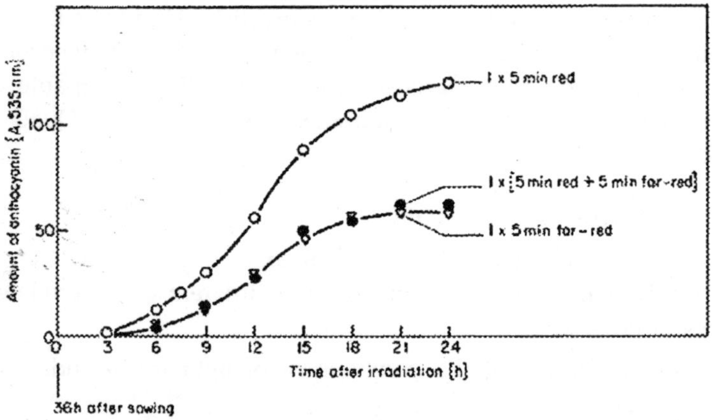

Figure: Accumulation of anthocyanin in* Sinapis *in the dark following irradiation treatments at time zero with 5 min red (o), 5 min far-red (Ñ), or 5 min red followed immediately by 5 min far-red (●) light (after Mohr* et al., *1971).

This simple red/far-red photoreversibility forms the basis of the concept that *Pfr* is the biologically active form of the pigment whereas *Pr* is inactive. Attempts to quantify the relationship between the number of *Pfr* molecules formed and the magnitude of the induced response have been both indirect and direct.

Indirect correlations are based on the premise that observed increases in the magnitude of the response with increasing light dose are a function of the degree of photoconversion of *Pr* to *Pfr*, i.e. the more quanta, the more *Pr* is converted to *Pfr* and therefore the greater is the response. The increase in anthocyanin in response to increasing doses of red light illustrates this point (Lange *et al.*, 1971). The so-called law of reciprocity (irradiance × time = constant) must hold for the light doses used for this interpretation to be valid. This establishes that the magnitude of the response is directly proportional to the total number of incident quanta regardless of the time or irradiance of the irradiation providing those quanta. The effects of irradiance level during the brief irradiations used in induction-reversion experiments are thus attributed entirely to the degree of photoconversion.

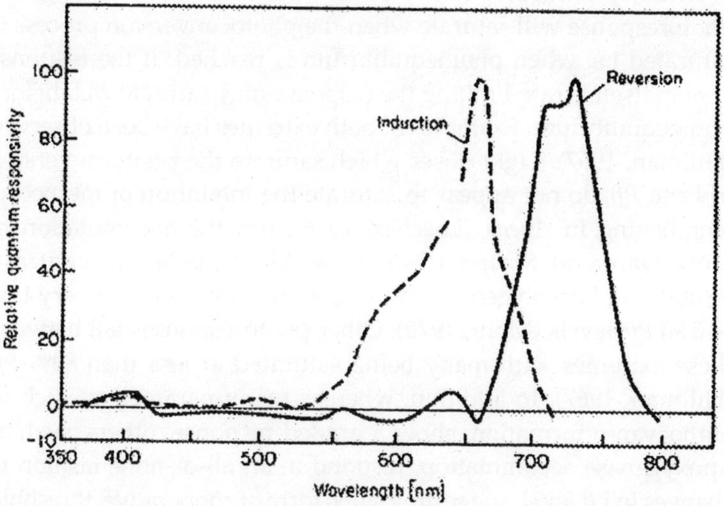

Figure: Action spectra for induction and reversion of plumular hook opening in bean seedlings (after Withrow* et al., *1957).

Biological action spectra are an extension of this principle. The magnitude of the response at different wave-lengths is interpreted to be a function of the relative effectiveness of the quanta at those wavelengths in the phytochrome photoconversion process. The close agreement between the action spectra of several biological responses on the one hand and those of the phototransformation reactions of the isolated pigment on the other lends strong support to this notion (Borthwick, 1972; Shropshire, 1972). In these cases a seemingly good correlation exists between *Pfr* level and response magnitude.

In contrast, however, the majority of rigorous attempts to demonstrate a direct quantitative correlation between the *photometrically* detectable *Pfr* level and the relevant biological response in the same system have been unsuccessful (Hillman, 1972). The reasons for this are not understood. An apparent exception is lipoxygenase levels in *Sinapis* (Oelze-Karrow & Mohr, 1973).

Increases in response with increasing doses of quanta must eventually saturate. If the level of *Pfr* is rate-limiting the photoresponse will saturate when the photoconversion process is saturated i.e. when photoequilibrium is reached. If the response system itself is rate-limiting the response may saturate well before photoequilibrium. Examples of both extremes have been observed (Hillman, 1967). Light doses which saturate the photoconversion of *Pr* to *Pfr* do not appear to saturate the inhibition of mesocotyl lengthening in *Avena* (Loecher, 1966), nor the accumulation of anthocyanin in *Sinapis* (Drumm & Mohr, 1974). In contrast, inhibition of lipoxygenase accumulation is saturated by very low (< 3%) *Pfr* levels (Mohr, 1972). Other photoresponses fall between these extremes with many being saturated at less than 80% *Pfr* (Hillman, 1967). In addition, whereas many parameters, such as anthocyanin formation, show a graded response, others, such as lipoxygenase accumulation, respond in an all-or-none fashion to changes in *Pfr* level, suggesting some form of cooperative, threshold mechanism (Oelze-Karrow & Mohr, 1973).

Implicit in the far-red reversibility of an induced response is that *Pfr* can act in the dark. Light is strictly a trigger. The magnitude and multiplicity of the responses indicate an extensive amplification mechanism. Unlike photosynthesis where light energy is converted stoichiometrically with quantum yields of less than 1.0, the low irradiances which actuate these phytochrome responses lead to final quantum yields well in excess of unity (Galston, 1974).

High Irradiance Responses (HIR)

If *Pfr* is, as postulated, the active, effector, it is clear that tripartite correlations between light dose, *Pfr* level and response magnitude will only be expected for irradiations terminated prior to photoequilibrium. Once photoequilibrium has been established, *Pfr* le vels are no longer irradiance dependent whether for short term or long term (Schäfer, 1975) irradiations. No further increase in the response should result therefore from further increases in irradiance after photoequilibrium.

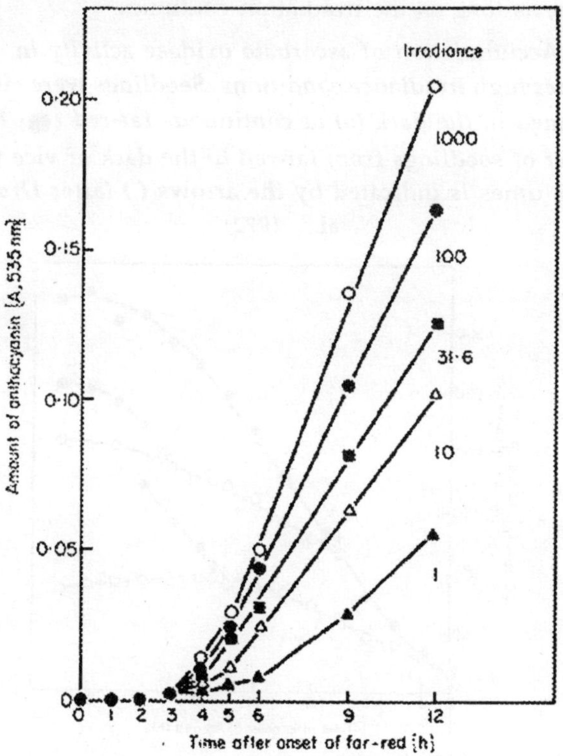

Figure: Accumulation of anthocyanin in Sinapis under high irradiance conditions. Seedlings were held in continuous far-red light of varying irradiance. All irradiances are expressed relative to the arbitrary value of 1,000 (= 350 μW. cm^{-2}) (after Lange et al., 1971).

Some parameters do, however, show a strong irradiance dependence at photoequilibrium. This effect is termed the high irradiance response (HIR) (Mohr, 1969). Such responses are observed with continuous irradiation where a phytochrome photosteady-state is rapidly established and maintained for prolonged periods.

The accumulation of anthocyanin in continuous far-red light illustrates this point (Lange *et al.*, 1971). The rate of accumulation is a function of the irradiance. This effect is only maintained, however, as long as the irradiation continues.

Figure: Accumulation of ascorbate oxidase activity in **Sinapis** ***under high irradiance conditions. Seedlings were either retained in the dark (o) or continuous far-red (●) light. Transfer of seedlings from far-red to the dark or vice versa at various times is indicated by the arrows ( ) (after Drumm*** **et al.,** ***1972).***

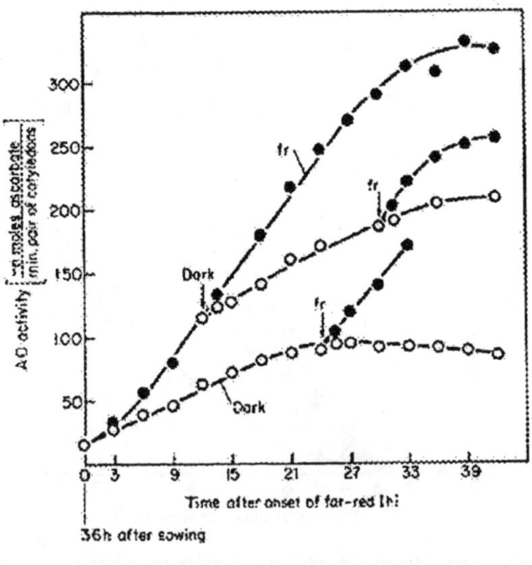

The irradiance-enhanced response rate reverts rapidly to that of the dark controls when irradiation ceases, and resumes again upon further irradiation. Reciprocity is not demonstrable for the HIR and, in some cases e.g. lettuce hypocotyl lengthening, the variable itself does not exhibit the classical red/far-red reversible response (Hartmann, 1966). Anthocyanin accumulation, in contrast exhibits *both* modes of response.

Action spectra of HIR have peaks in the blue and far-red (Borthwick *et al.*, 1969; Mohr, 1969). The most extensively studied response is that of the inhibition of lettuce hypocotyl lengthening (Hartmann, 1967). The seedlings were irradiated continuously for 18 hours with monochromatic light of different wavelengths and irradiances. As with other HIR there is no coincidence of the observed spectrum with the absorption maxima of either *Pr* or *Pfr*.

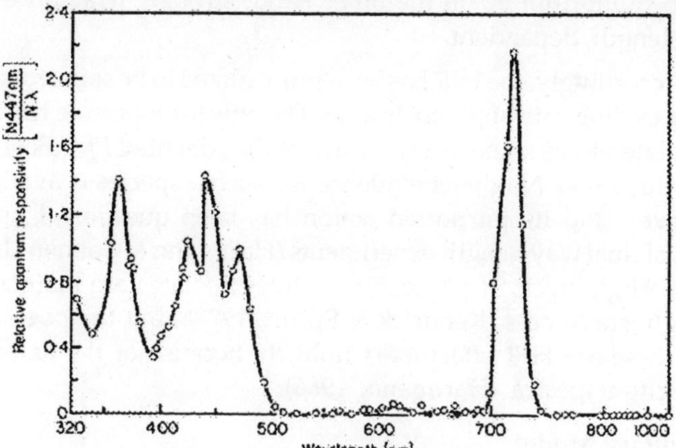

Figure: Action spectrum for inhibition of lettuce hypocotyl lengthening under continuous irradiation (after Hartmann, 1967).

The conclusion that phytochrome mediates the high irradiance effects derives from other data using the same plant system (Hartmann, 1966). Prolonged irradiations with two wavelengths which were relatively ineffective when given separately (658 nm and 766 nm) were highly effective when given simultaneously. The maximum effect with these and other wavelength pairs always occurred where the photosteady-state *Pfr* concentration was about 3%. This agrees well with the peak of activity observed at 720 nm with single wavelength monochromatic light also known to establish about 3% *Pfr* at photoequilibrium. Furthermore, the effectiveness of a single wavelength irradiation at 717 nm could be nullified by simultaneous irradiations of either 658 nm or 759 nm. These wavelengths would shift the photoequilibrium away from 3% *Pfr* towards higher or lower values respectively.

The question remains, however, as to *how* both the irradiance and wavelength dependency of the HIR can be explained in terms of phytochrome. The response is unlikely to be a function of the absolute *Pfr* level as this is irradiance and wavelength *independent* with prolonged irradiations (Schäfer, 1975). The rate at which the pigment molecule oscillates between the two forms at photoequilibrium is, on the other hand, strongly irradiance and wavelength dependent.

Accordingly, the HIR has been rationalized to be some function of the cycling rate of phytochrome. The effector molecule has been postulated to be some 'excited form of *Pfr*', denoted *Pfr** (Schopfer & Mohr, 1972). No direct evidence for such a species is available, however, and its purported action has been questioned on the basis of dual wavelength experiments (Hartmann & Cohnen Unser, 1973). High levels of phytochrome intermediates are maintained at high irradiances (Kendrick & Spruit, 1973a) but the possibility that these are HIR effectors is unlikely because of the nature of the action spectra (Hartmann, 1966).

A Unitary Model

Schäfer (1975) has recently developed a single, formalistic model which can theoretically account for both induction-reversion and HIR responses, whether graded or cooperative, in terms of phytochrome. The pigment is postulated to be a bimodal ligand with the forms *Pr* and *Pfr* which interact with receptor sites also having dual forms, *X* and *X'*:

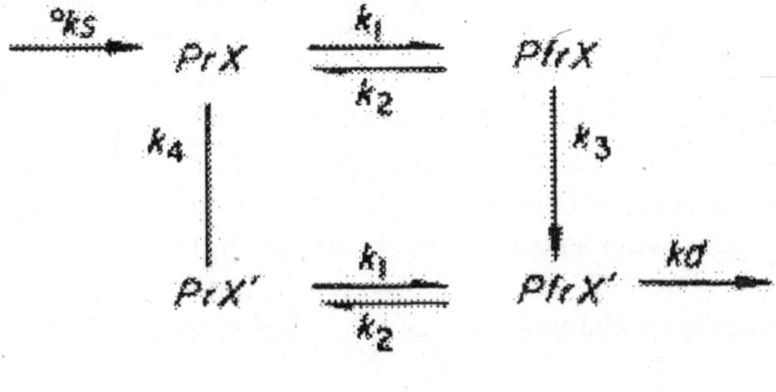

where °*ks* and *kd* are the rate constants for synthesis and destruction of phytochrome respectively; k_1 and k_2 rate constants for the photoconversion reactions; and k_3 and k_4 rate constants for the *X* ® *X′* and *X′* ® *X* transitions.

The irradiance and wavelength dependencies of the HIR response then become explicable in terms of the flux rates through the cycle under continuous irradiation.

The basic conclusions reached are that *PfrX′* is the effector element in induction-reversion responses, whereas *PfrX* or the flux k_3. *PfrX* is the effector element in HIR. The molecular nature of X and the mechanism of action of the effector elements are, however, unresolved questions.

Modes of Light Signal Transmission

Given that phytochrome mediates both induction-reversion and HIR, it is clear that the photoreceptor utilizes the incoming light signal differently in each case. Whereas induction-reversion responses are induced by a red pulse and can develop in the subsequent dark period, HIR require a continuous light energy input to sustain the response. Increases in irradiance lead to increases in response magnitude in both cases but the effect is interpreted differently for each.

For induction-reversion responses the irradiance effect is only observed prior to photoequilibrium; is interchangeable with time of irradiation (reciprocity holds); and is interpreted as reflecting the effectiveness of the total light dose in determining the degree of photoconversion of *Pr* to *Pfr*. *Pfr* is considered the effector molecule and some form of *Pfr* -response stoichiometry is expected.

For HIR, on the other hand, the irradiance effect is observed at photoequilibrium; is not interchangeable with time of irradiation (reciprocity does not hold); and is interpreted as being some function of the phytochrome cycling rate.

The precise effector molecule or process is uncertain. For induction-reversion responses light is viewed simply as a trigger and as having no further direct role in the inductive function of *Pfr*. The light signal is 'stored' in the *Pfr* form for subsequent utilization. HIR, in contrast, require a sustained, direct interaction of the photoreceptor with the incident excitation energy.

This indicates that these effects are light-driven as distinct from being light-triggered. Light appears to have a direct role in the inductive function of the pigment, the energy input being rapidly dissipated.

The suggestion that photosynthesis or cyclic photophosphorylation might in some way be responsible for HIR has been advanced but several pieces of evidence argue against this (Mohr, 1972). Speculative suggestions that phytochrome might function as a photocoupler (Quail, 1975a) or a specific, lightdriven permease (Smith, 1970) in the HIR have also been advanced but not substantiated.

Response Kinetics

Phytochrome is considered to trigger, in some way, a chain of events leading sooner or later to a measurable biological response. The initial triggering of those processes necessary for the development of the response can be termed phytochrome 'action'; and the appearance of a measurable change in the parameter being monitored can be called response 'expression'.

This leads to the recognition of three categories of phenomena: (a) rapid action/rapid expression; (b) rapid action/delayed expression; and (c) delayed action/delayed expression responses. 'Rapid' here arbitrarily means 10 minutes and 'delayed' ^{3} 30 minutes after the initial photoconversion act.

Rapid Action/Rapid Expression Responses

As phytochrome action must either coincide with or precede expression, rapid action can be implied from the kinetics of the expression alone in these cases. A red/far-red reversible change in electric potential in *Avena* coleoptiles within 15 seconds of the start of irradiation is the most rapid phytochrome-mediated phenomenon thus far reported (Newman & Briggs, 1972).

Similar changes have been observed in the biolelectric potential of mung bean root tips exposed to successive red and far-red irradiations (Jaffe, 1968; Racusen & Miller, 1972). Such changes had earlier been inferred from the red/far-red reversible adhesion of root tips to negatively charged glass surfaces (Tanada, 1968).

Both responses are detectable within 30 seconds of the start of irradiation. These changes in surface charge are interpreted to indicate changes in the plasmalemma. Fluorescent probe studies support this notion (Racusen, 1973). Red light also induces H^+ efflux from root tips (Yunghans & Jaffe, 1972).

Phytochrome regulates leaflet movement in *Mimosa, Albizia* and *Samanea* (Satter & Galston, 1975). This movement is accompanied by an energy-dependent transfer of K^+ ions between the ventral and dorsal motor cells of the pulvinus. Both effects are detectable 10 minutes after red or far-red irradiations.

The changes in K^+ are correlated with changes in transmembrane potential (Racusen & Satter, 1974). Furthermore, red/far-red regulated changes in surface charge similar to those of root tips are detectable 30 to 120 seconds after irradiation.

Figure: Kinetics of (A) root tip adhesion to a negatively charged glass surface and (B) the development of a biolectric potential across the root tip in response to irradiation with red or far-red light (after Jaffe, 1968).

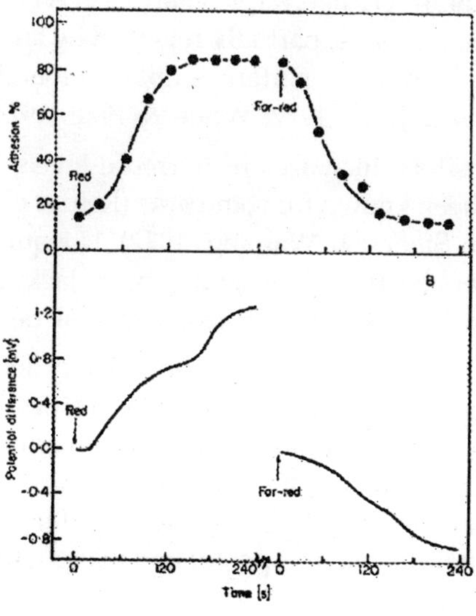

These effects are also strongly indicative of phytochrome-mediated changes in plasmamembrane properties. Phytochrome-regulated changes in the rate of plasmolysis have been detected in *Mougeotia* within 6 minutes of the onset of red irradiations (Wiesenseel & Smeibidl, 1973). A change in plastid orientation is also evident in this alga in less than 10 minutes after a red pulse (Haupt, 1972a). The latter effect can only be fully reversed by far-red during the first minute after red light.

Thus potentiation of the response has begun within 1 minute of photoconversion. Available data indicate that the effective phytochrome is located on or near the plasmalemma. Contractile fibrils appear to be responsible for the chloroplast movement (Schönbohm, 1973).

Red light (15 seconds) induces an increase in growth rate in coleoptile tips within 60 seconds of the start of irradiation (Weintraub & Lawson, 1972).

The effect is partially reversed if far-red immediately follows the red. Inhibitors of transcription and translation are without effect.

As the cell must regulate wall extension through the plasmalemma, this response might also represent a phytochrome-mediated membrane change. Rapid changes in ATP levels, induced by red and, in some cases, partially reversed by far-red, have been reported but no consistent pattern is obvious (Sandmeier & Ivart, 1972; Yunghans & Jaffe, 1972; White & Pike, 1974).

Rapid transitory increases in hormore levels in response to red light have been known for some time (Reid *et al.*, 1968; Beevers *et al.*, 1970; van Staden & Wareing, 1972). Unequivocal evidence of far-red reversibility has generally been lacking in the past, however, often because of poor experimental design.

Recently, the level of gibberellin activity extractable from etioplast-rich preparations from grass leaves has been shown to increase 3-fold within 5 minutes of the termination of a 5 minute red irradiation of the isolated fraction (Evans, 1975; Evans & Smith, 1976a). This effect is reversed by far-red light given immediately after the red. It is postulated that phytochrome causes the movement of gibberellin across the etioplast envelope into the ambient medium.

These observations, and those made with mitochondria-rich fractions (Manabe & Furuya, 1974), are rapid, of potential physiological significance, and suggestive of *in vitro* phytochrome-mediated changes in the functional properties of membranes in the isolated fractions.

Lipoxygenase accumulation in *Sinapis* responds to changes in *Pfr* level in less than 5 minutes under appropriate conditions (Oelze-Karrow & Mohr, 1973). It has been concluded that the level of the enzyme in the cotyledons is controlled by phytochrome in the hypocotyl hook, through a highly cooperative threshold mechanism that responds to, and is saturated by, *Pfr* levels of 1–2% of the total pigment (Oelze-Karrow & Mohr, 1974).

Other rapid, inter-organ transfer of phytochrome signals have also been reported (de Greef & Caubergs, 1973). Some form of biophysical transmission system has been postulated, with the membrane continuum of the plasmodesmata as a suggested candidate (Oelze-Karrow & Mohr, 1973).

Rapid Action/Delayed Expression Responses

Rapid phytochrome action in these cases is deduced from the rate at which the response escapes susceptibility to photoreversal by far-red light following an inductive red pulse.

The actual response may not be expressed for days after the irradiation although the inevitability of its appearance has long since been irreversibly established. *Pfr* is said to have 'potentiated' the response (Borthwick, 1972). The escape from reversibility is viewed as the *Pfr* -triggered reaction chain having progressed beyond those steps directly under phytochrome confrol.

The effect of a red light pulse on flowering in *Pharbitis* is only partially reversed by far-red given 30 seconds after the start of the red irradiation (Fredericq, 1964). After 3 minutes, far-red no longer reverses the effect. Thus although the flowering response itself is not expressed for several days, it is potentiated within seconds by *Pfr*.

Flowering in *Chenopodium album* and *Kalanchoe* behaves similarly. A synergism between phytochrome and gibberellin in lettuce seed germination (Bewley *et al.*, 1968) and the de-etiolation response of *Pisum* (Haupt, 1972a) are also in this category.

Delayed Action/Delayed Expression Responses

These responses show a lag from irradiation to expression but are readily reversed by far-red over relatively long periods in the dark after the red pulse. Gradual escape from reversibility can occur but is relatively slow. This is viewed as indicating that the continued presence of *Pfr* is required over a relatively long period in the dark to maximise expression.

An important feature of these responses is that dark reversion and destruction are continually and often rapidly depleting the *Pfr* pool during this period.

The vast majority of recorded red/far-red reversible responses, too numerous to catalogue, are included in this category (see Mitrakos & Shropshire, 1972; Mohr, 1972; Smith, 1975).

Lettuce seed irradiated for I minute with red light germinates up to 100% 24 hours later. Far-red reverses this effect up to 12 hours after the red irradiation but with decreasing effectiveness. Anthocyanin formation in *Sinapis* has a lag of 3 hours from irradiation to the onset of accumulation.

During this time the effect becomes decreasingly susceptible to reversal by far-red but escape is never complete. The *in vitro* protein synthetic capacity of 80s ribosomes from corn is enhanced by 5 minutes of red light prior to harvest (Travis *et al.*, 1974). This effect is detectable within 30 minutes of red irradiation and escapes far-red reversibility within an hour.

Response Manifestations

Phytochrome has a regulatory role in all major phases of plant growth and development. The changes in cellular biochemistry and physiology which underlie this regulation are detectable at almost any chosen level. Some responses are biophysical in nature (membrane potential), others biochemical (enzyme levels); some require protein synthesis (leaf expansion), others do not (leaf movements); some involve cell division (seed germination), others only cell expansion (plumular hook opening); some are restricted to the irradiated cells (leaf movements), others result from transmissable stimuli (floral induction); some are dependent on phytochrome-endogenous rhythm interactions (flowering), others appear independent (seed germination) (Galston, 1974; Satter & Galston, 1975).

When the nature of the responses is coupled with their kinetic properties a pattern emerges. In general, the most rapid responses are surface or membrane-associated phenomena, often physico-chemical in nature and independent of RNA and protein synthesis. Other cellular processes such as changes in enzyme levels mostly respond more slowly.

MECHANISM OF ACTION

The molecular mechanism of phytochrome action is the central issue in phytochrome research. The term 'primary reaction of phytochrome' has been used to describe the first reaction in which the physiologically active species becomes involved (Mohr, 1972). Assuming *Pfr* to be the active species this reaction has been formalized:

$$\text{Pfr} + \text{X} \rightarrow \text{PfrX}$$

where X satisfies the logical necessity of a reaction partner regardless of its nature. The 'primary *action*' of phytochrome then is presumably the induction of a functional *change* in the reaction partner ($X \rightarrow X'$). These are useful formalistic treatments but say nothing of the actual molecular mechanisms involved.

Any molecular hypothesis of phytochrome action must provide an explanation for the specificity, multiplicity and amplification of the induced responses, and attempt to reconcile the seemingly disparate modes of light signal transmission in induction-reversion and high irradiance responses.

Specificity here refers to the observed capacity of different plants, organs, adjacent cells in the same tissue and even different processes in the same cell to respond differently to the same effector molecule (Mohr, 1972). All major hypotheses hold the view that this property resides in the fundamental programming of the target cells themselves rather than in the phytochrome molecule.

Multiplicity refers to the vast diversity in form, direction, magnitude and timing of response expression *per se*. This can be accounted for by a suitable amplification mechanism. Two major hypotheses of phytochrome action, based on this rationale, have been advanced in recent years: the differential gene activation hypothesis (Mohr, 1966) and the membrane permeability hypothesis (Hendricks & Borthwick, 1967).

The possibilities that plant hormones (Black & Vlitos, 1972) or acetylcholine (Jaffe, 1970) might be primary mediators of phytochrome effects have also been investigated but convincing supporting data are sparse (Tanada, 1972).

The gene activation hypothesis in its broadest sense is trivial, as gross morphogenetic changes such as flowering are unlikely to occur without some modification of gene expression.

In its strictest sense the hypothesis implies that the primary action of phytochrome is to alter genetic activity. This requires direct *Pfr* -genome interaction. Data relevant to this problem come mainly from measurements of phytochrome-induced changes in enzyme activities and related phenomena (Mohr, 1972).

There are two major criticisms of this hypothesis. First, while the data so far obtained are consistent with the general framework of the concept, they provide no evidence that *Pfr* interacts directly with the genome.

Increased incorporation of [-H]-uridine into high molecular weight RNA in continuous far-red light has been reported and interpreted as phytochrome-enhanced transcription of rRNA cistrons (Thien & Schopfer, 1975).

Phytochrome-mediated *de novo* synthesis of enzyme proteins has also been demonstrated (Attridge, 1974; Acton & Schopfer, 1974) but this does not establish that control is at the gene level. Indeed, insensitivity of responses such as triose phosphate dehydrogenase (Cerff, 1974) and ascorbate (Schopfer, 1967) accumulation towards actinomycin D might suggest otherwise in these cases.

Evidence of phytochrome-mediated activation of precursors of phenylalanine ammonia-lyase has been advanced (Attridge *et al.*, 1974) but this point is controversial (Acton & Shopfer, 1975).

The second criticism is that a growing list of responses appear unlikely to be explained by changes in gene expression. These are in general the rapid action/rapid expression responses.

These are more rapid than would be expected of gene regulation and are often insensitive to inhibitors of transcription and translation (Satter & Galston, 1973). Furthermore, lipoxygenase, the only enzyme responding rapidly enough to be placed in this category, is now postu-lated to be controlled by phytochrome in a separate organ (Oelze-Karrow & Mohr, 1974).

The membrane hypothesis postulates that phytochrome interacts directly with cellular membranes i.e. the primary action of *Pfr* is to modify membrane properties. Indirect support for this notion derives from the fact that the majority of the most rapid responses are either surface phenomena or can be rationalized in terms of membrane changes.

Direct evidence of an association of phytochrome with an identifiable cellular membrane is yet to be advanced however although Evans and Smith (1976b) have recently reported the presence of phytochrome in fractions enriched for etioplast envelope membranes.

A postulated membrane locale of phytochrome action permits the formulation of a unitary hypothesis to account for all phytochrome-mediated phenomena. Both rapid and long term effects can be rationalized to have emanated from the same primary event.

A single, fundamental alteration in membrane properties, perhaps simultaneously in several cellular membranes, provides an immediate mechanism for the generation and amplification of a multiplicity of secondary effects (Changeux, 1969).

Such effects might include changes in ion flux, activation of membrane bound enzymes, altered compartmentalization and soon. Any or all of these might lead directly or indirectly to altered gene expression. Moreover, a phytochrome-membrane association suggests a possible basis for the dual function of the pigment in induction reversion and HIR responses.

The opportunity would exist for the molecule to function either as a reversible, stereospecific membrane effector, *via* photoinduced changes in its protein conformation; and/or as a photocoupler (Quail, 1975a) or light driven permease (Smith, 1970) *via* pigment cycling.

The speculative nature of such suggestions, however, is evidence of our ignorance in this area. Indeed the basic concept of only a *single*, primary phytochrome reaction has been challenged (Mohr, 1974).

It has been suggested instead that a multiplicity of reaction partners and corresponding primary reactions must be postulated to account for both graded and cooperative responses in the same system (Mohr *et al.*, 1971). This would circumvent the second criticism of the gene regulation hypothesis and obviate any potential conflict between the two major hypotheses. The molecular mechanism of phytochrome action is clearly an open question.

5

PLANT CELLS MANIPULATION

INTRODUCTION

The ultimate objectives of any scientific investigation are twofold–to increase understanding, and to create opportunities for the application of the new knowledge, ostensibly in the service of man. The two objectives may not often be consciously connected in the minds of the investigators, but history and experience have demonstrated that new information always stimulates a drive towards practical application. This is already beginning to be true for plant cell biology through the directed manipulation of plant cells grown in culture.

As is described, it has been possible to culture organs, tissues and cells of higher plants under aseptic conditions for many years. These techniques, developed largely through assiduous application of the trial-and-error approach, are now becoming highly sophisticated and have already yielded information of great value to our understanding of cellular processes, particularly the cell cycle. In recent years, it has become possible to remove the walls from plant cells by enzymatic procedures, yielding naked protoplasts, some of which remain viable and may be propagated further.

The exploitation of cell and protoplast culture techniques for purely scientific purposes is still in its infancy, but already the methods have been used commercially for the propagation of valuable stock plants, and for the propagation of virus-free plants from virus-infected stocks. What makes cultured cells and protoplasts so exciting from the practical viewpoint, however, is the potential they appear to offer for artificial genetic manipulation. It is too early, yet, to say whether protoplast fusion or gene transfers will yield new and revolutionary varieties of crop plants—the prospect does exist however, and a consideration of these matters seems a fitting note on which to conclude our coverage of the molecular biology of plant cells.

PLANT CELL CULTURE

Almost every chapter of this book illustrates that research in plant cell physiology and biochemistry is, at each moment in time, limited by the resolving power of current experimental techniques and by the availability of appropriate experimental plant material. Here we are concerned with the development of new experimental materials (cultured organs, tissues and cells), with culture systems being evolved for their exploitation and with recent experimental work which illustrates that these culture systems are adding a new dimension to studies in plant cell physiology.

The complexity of higher plants has inevitably led workers, concerned with investigating particular aspects of plant physiology, to the use of systems which are of reduced complexity. Hence the early and continuing use by plant physiologists of isolated plant organs (e.g. seedling roots, leaves), complex tissue systems (e.g. discs cut from leaves and storage organs) organ segments (e.g. segments of coleoptiles and hypocotyls) and isolated cell organelles. However, isolated organs and organ fragments are still systems of very considerable complexity and they are, from the beginning, systems of declining viability and favourable sites for colonization by microorganisms.

The concept that the aseptic *culture* of isolated organs, tissues and cells would 'give some interesting insight into the properties and potentialities which the cell as an elementary organism possesses' and 'would provide information about the inter-relationships and complementary influences to which cells within the multicellular whole organism are exposed' was formulated as early as 1902 by Haberlandt. Then, after a lapse of more than 30 years, White (1934) described successful root cultures initiated from the root tips of tomato seedlings, and White (1939) and Gautheret (1939) demonstrated that the parenchymatous wound callus which frequently forms at the exposed surfaces of organ segments could be removed and grown indefinitely as a relatively undifferentiated tissue (callus) culture. From this period there has been rapid progress in organ and tissue culture. Root cultures have been developed from many species (Butcher & Street, 1964). Cultures of stem apices (meristem culture) and of leaf, flower and fruit primordia have been successfully established (Street, 1969).

These organ cultures as experimental systems differ from isolated organs in two important respects; they are sterile (free from microorganism contamination) and are handled aseptically, and they are sytems where unimpaired viability is evidenced by growth involving both cell division and cell expansion. Such cultures have contributed to our knowledge of the specific nutritional and hormonal requirements essential for the growth and development of the separate organs of the whole plant and of their specific physiology and biosynthetic activities (Street, 1969). Callus cultures have also now been established from a very wide range of species and have been used in studies on the initiation of root and shoot primordia (Street, 1975a), on cytodifferentiation (Wetmore & Rier, 1963, Gautheret, 1966; Torrey, 1971), on the induction of division in quiescent tissue cells (Yeomann & Aitchison, 1973), on the nature of plant tumour cells (Butcher, 1973) and on the synthesis of a diversity of secondary plant products (Yeomann & Aitchison, 1973).

In 1953, Muir reported that if fragments of callus cultures of *Tagetes erecta* or *Nicotiana tabacum* were transferred to liquid medium and agitated on a reciprocal shaker, then the callus fragments broke up to give a suspension of single cells and small aggregates of cells and that this suspension contained actively dividing cells and hence could be propagated by serial subculture (Muir, Hilderbrandt & Riker, 1954). Such liquid cell suspension cultures have now been obtained from calluses of a number of species and this chapter will outline the development of more sophisticated techniques for their culture and assess their value for studies on the control of growth, metabolism and differentiation in higher plant cells.

CHANGES IN GROWTH AND METABOLISM OF PLANT CELLS IN BATCH CULTURE-CYTODIFFERENTIATION

Plant cells in batch culture, i.e. cultures in a fixed volume of culture medium, increase in biomass by cell division and cell growth until a factor in the culture environment becomes limiting and sends them into a stationary phase. When such stationary phase cells are subcultured they pass in succession through a lag phase, a short-lived period of exponential growth, a period of declining relative growth rate and then again enter stationary phase. Traditionally such cultures are initiated by an inoculum establishing a relatively high initial cell density and only accomplish a very limited number of divisions before entering stationary phase. For example cell cultures of sycamore (*Acer pseudoplatanus*) initiated at *ca.* 2×10^5 cells ml^{-1} will reach a final cell density of *ca.* 3×10^6 cells ml^{-1} corresponding to 4 successive doublings of the initial population.

The degree of cellular aggregation in these cell cultures depends upon the species of cell, or cell line within a species, and the culture conditions, but always shows a basically similar pattern of change during the growth cycle of a batch culture. The culture at stationary phase contains the highest proportion of free cells and mean cell volume is at its maximum value. When subcultured to new medium, the cells first embark upon a massive synthesis of new cytoplasm and associated organelles and then begin to divide. For a short time cell division proceeds at a specific growth rate (μ) which is constant and maximal (μ_{max}) for that culture environment. During this phase mean cell volume declines sharply and the proportion of cells in aggregates rises. Then the specific growth rate begins to decline (slowly at first and later at an ever increasing rate) mean cell volume increases and, associated with this cell expansion, the aggregates break up and release free cells.

Associated with these growth and structural changes are changes in physiological activity. Measurements, on a per cell basis, of respiration, of the levels of individual cell constituents and of the activities of individual enzymes show that peaks of activity occur. These may be quite sharp and are not coincidental. Different metabolic patterns emerge and decline during the progress of batch culture. Thus RNA synthesis is initiated prior to cell division, proceeds for a time at a greater rate than cell number increase and then ceases whilst cell division is still proceeding (Short, Brown & Street, 1969; Nash & Davies, 1972). Free nucleotides (mainly UDP-glucose and ATP in sycamore cells) are synthesized rapidly during lag phase, presumably an essential preparation for subsequent synthesis of cell-wall polysaccharides and as an energy source for the endergonic processes of cell division, but their net synthesis ceases very shortly after the onset of division (Brown & Short, 1969).

Similar transient high activity in carbohydrate oxidation by the pentose phosphate pathway during lag phase has been interpreted as providing the necessary NADPH for the massive biosynthesis achieved during the lag phase of the growth cycle (Fowler, 1971).

By contrast, the very sharp peak in ethylene production in sycamore cell cultures occurs late in the cell division phase when the cells are beginning to increase in mean cell volume and may be responsible for initiating aggregate breakdown (Mackenzie & Street, 1970).

An essentially similar pattern of ethylene production has been reported for cell cultures of *Rosa* spp., *Glycine max, Triticum monococcum, Melilotus alba, Haplopappus gracilis* and *Ruta graveolens* (La Rue & Gamborg, 1971). Large changes in the activity of phenylalanine ammonia-lyase (PAL) and in *p* -coumarate: CoA ligase occur prior to stationary phase in cultures of *Glycine max* (Hahlbrock, Kühlen & Lindl, 1971).

A similar peak of PAL activity has been reported in cell cultures of *Rosa* sp. (Davies, 1972). These changes coincide with maximum production of total phenols by the cultures. Other secondary products are produced by cell cultures, the time of maximum synthesis being restricted to a phase in the growth cycle and often being markedly influenced in its intensity by the plant growth hormone composition of the culture medium (e.g. hemicelluloses and lignin by sycamore; see Carcellar, Davey, Fowler & Street, 1971; visnagin (a physiologically active furanochromone) by *Ammi visnaga;* see Kaul & Staba, 1967; caffeine by tea cell cultures; see Ogutuga & Northcote, 1970; various alkaloids, by cell cultures of solanaceous plants; see Tabata, Yamanito & Hiroaka, 1971).

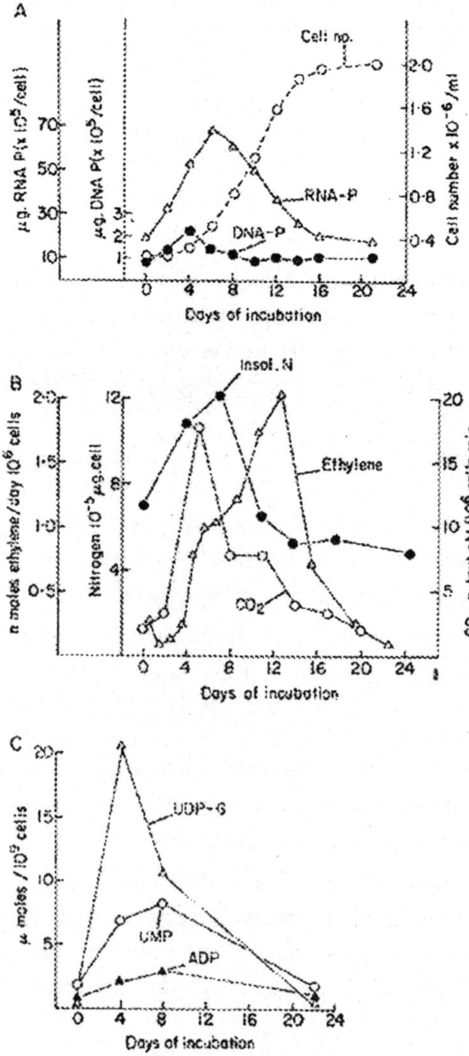

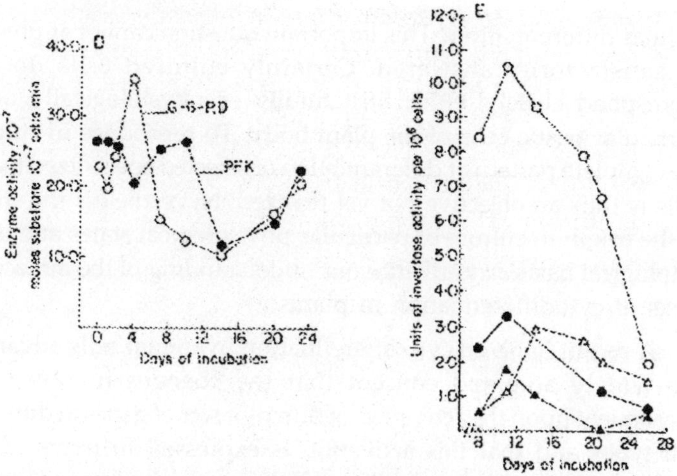

Figure: Changes in metabolic activity during the progress of growth of sycamore (Acer pseudoplatanus L.) cells in batch culture. (a) Growth curve and data for total DNA and RNA per cell (from Short, Brown & Street, 1969.) (b) Data for total cellular nitrogen and respiration (from Givan & Collin, 1969) and for ethylene production (from MacKenzie & Street, 1970.) (c) Data for levels of nucleotides (ADP=adenosine diphosphate, UMP=uridine monophosphate, UDP-G= uridine diphosphate glucose) (from Brown & Short, 1969.) (d) Data for activities of glucose-6-phosphate dehydrogenase (G-6-PD) and phosphofuctokinase (PFK) (from Fowler, 1971). (e) Data for cell wall invertase assayed at pH 4.5 ( ) and for soluble invertase assayed at pH 4.5 and pH 7.0 (from Copping & Street, 1972.)

Thus during the progress of batch culture the cells pass through a series of contrasted physiological states which encompass cells becoming meristematic, cells expressing high meristematic activity, cells undergoing expansion and becoming either metabolically quiescent or in which certain restricted metabolic pathways are emphasized.

Are these large changes in cellular structure and metabolic activity observed in batch-propagated cell cultures examples of cellular differentiation? This important question cannot at present be satisfactorily answered. Certainly cultured cells do not correspond closely, either structurally or physiologically, with particular tissue cells of the plant body. To reproduce in culture the complete pattern of differentiation of selected specialized tissue cells is thus an objective not yet realized. Nevertheless the study of the origin in culture of particular physiological states and their cytological basis *may* advance our understanding of the molecular basis of cytodifferentiation in plants.

A recent book on cytodifferentiation in animal cells advances the widely accepted concept that the changes involved are consequent upon the activation of different sets of genes in different cell types and that this activation is expressed in terms of the synthesis of enzymes and other cellular proteins (Truman, 1974). Further, the differentiation process is regarded as a relatively permanent and irreversible change. In support of this it is possible to quote studies such as those of Cahn and Cahn (1966) on the culture of retina cells; such cells continue to produce the characteristic pigment granules during prolonged culture under conditions conducive to active cell division whereas *in vivo* such cells are non-dividing.

If under certain secondary conditions of culture pigment granules were lost, nevertheless when the cells were returned to the primary conditions of culture the pigmentation returned. To distinguish more minor and readily reversible changes in physiology from cytodifferentiation the former are described as modulations and are considered to reflect the operation of allosteric and other 'fine' processes of metabolic regulation. However, Truman (1974) concedes that 'it is not at all easy to draw a distinction between those enzymes which are fundamental to differentiation and those which represent very short-term modulations in the activity of cells' and in considering cytodifferentiation in liver cells states that 'differentiation and modulation do not represent distinct processes but are merely the extreme ends of a spectrum of changes that can occur'.

Although from a number of plant species, cell cultures can be readily initiated from different organs (roots, stems, leaf petiole or lamina, cotyledons etc.) or from different living tissues within an organ (parenchymatous cells of pith, cortex, or mesophyll, cambial and other meristematic cells, immature vascular cells etc.) there is no very convincing evidence that they retain, in culture, characteristics of their *in vivo* origin although, as will be mentioned later in this chapter, they may not have undergone, during culture induction, the required degree of dedifferentiation necessary to express their totipotency (i.e. the capacity to generate a new plant in the way normally achieved from the fertilized egg). Such observations suggest that cytodifferentiation in higher plants, provided it has not proceeded to the point where cell death is inevitable, is a more readily reversible process than in the cells of higher animals.

Hence the readily reversible physiological states observed in plant cell cultures may be basically identical with the states involved in normal cytodifferentiation. Further, it raises the possibility that cytodifferentiation does not depend on the transcriptional activity of different sets of genes for each kind of tissue cell but that the 'specialized' physiology of such cells may represent the influence of cytoplasmic factors (plant hormones?) on the stability and transport of RNA species and other aspects of the translational steps in gene expression. As discussed below, it may be possible to examine such hypotheses experimentally with plant cell cultures.

STEADY STATES OF GROWTH AND METABOLISM OF PLANT CELLS IN CONTINUOUS CULTURE

Reference was made above to the short period of exponential growth observed in batch cultures.

However, even during this phase of the growth cycle, the cells do not achieve a steady state (a state of balanced growth); cell division is uncoupled from increase in cell dry weight and protein content so that the cells are changing in size and composition despite the constancy of the double time of the culture. Such observations raised the question of whether balanced growth could be achieved in plant cell cultures if a constant culture environment could be established by developing open continuous culture systems (systems in which inflow of fresh medium is balanced by outflow of an equal volume of culture). Wilson, King & Street (1971) have developed such a system providing conditions of aeration and agitation appropriate to plant cell cultures, capable of long-term aseptic operation and functional either as a chemostat or as a turbidostat.

Figure: Unbalanced growth of sycamore cells* (Acer pseudoplatanus *L) during the transient exponential growth phase achieved in batch culture. (a) Semilogarithmic plots showing rate of change in cell number, total protein and cell dry weight per unit volume of culture. The slopes of the lines of best fit (calculated by linear regression analysis, P < 0.01) were used to determine the specific growth rates (μ) for each parameter.

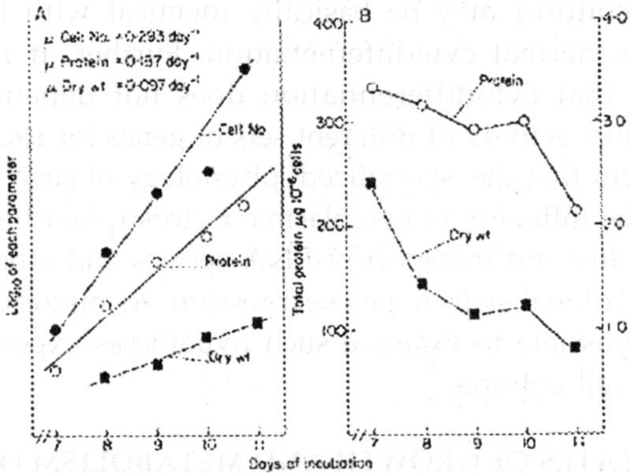

Chemostat Cultures

Subsequent work (King & Street, 1973; King Mansfield & Street, 1973) with this system operated as a chemostat (where equilibrium is established at a fixed rate of imput of a growth-limiting nutrient) has shown that long-term steady states of growth can be achieved with plant cell cultures and that such cultures conform to the chemostat theory developed from work with microorganisms (Monod, 1950; Novick & Sziland, 1950). In a chemostat, the relationship between cell density, x (cells per unit volume of culture), dilution rate, D (volume of new medium added per unit time expressed as a fraction of the total culture volume), specific growth rate, μ (increase in biomass per unit biomass per unit time) and time (t) is given by the equation

$$dx/dt = \mu x - Dx$$

When equilibrium is reached and a steady state established $dx/dt = 0$, $\mu = D$ and x has a value characteristic of the dilution rate. Further, the nutritive environment remains constant, each nutrient achieving an equilibrium concentration which is related to its imput concentration and the rate of its consumption by the culture. As defined above, the equilibrium achieved in a chemostat culture results from one particular nutrient (depending upon the composition of the culture medium) becoming the limiting nutrient and determining the specific growth rate (μ) of the cells.

The nature of these steady states can be illustrated from work involving sycamore cell cultures growing in a synthetic medium (Stuart & Street, 1969) in which the supply of nitrogen is the limiting factor. Data for one such steady state ($D = 0.194\ day^{-1}$). This shows that the cells are in a balanced state of growth and metabolism (as illustrated by the values for cell number, packed cell volume, cell dry weight, protein, DNA and RNA, and oxygen demand) and that the nutrient medium within the culture vessel is constant in composition (as illustrated by constancy of culture pH and the levels of glucose, phosphate and nitrate). Such steady state cells also display constant levels of metabolites (e.g. amino acids; Street, Gould & King, 1975) and constant levels of activity of individual enzymes (e.g. enzymes concerned with carbohydrate respiration; Fowler & Clifton, 1974; and with nitrogen assimilation; Young, 1973).

Such chemostat cultures can be operated from very low growth rates (i.e. low dilution rates) to growth rates approaching the maximum growth rate (μ_{max}) for the culture medium chosen, provided dilution rate is such that the cells can still achieve a matching growth rate. Of course if dilution rate is further increased, cell density does not stabilize and the culture suffers wash-out. Cells in balanced growth but highly contrasted in growth rate (and hence in cytology and metabolism) can therefore be obtained by chemostat culture. The range of change in certain cell parameters in sycamore cell cultures at different dilution rates (and hence specific growth rates) over the range D = 0.06–0.236 day^{-1} (corresponding to double times over the range 280–70 hr). What this means is that it is possible to stabilize at will, by fixing dilution rate at an appropriate level, the individual physiological states which have only a transient existence in batch culture.

Work with chemostat cultures has shown that the same cell population can be taken through a series of steady states and then if the dilution rate is returned to that of an earlier steady state the cells again achieve not only the new predictable growth rate but also the physiological activities earlier recorded as characteristic. Such experiments demonstate the full reversibility of the cytological and physiological changes invoked. Interest will now focus on detailed studies of the kinetics of the transition between steady states in terms of enzyme activities and metabolite levels. Preliminary studies along these lines have already shown that during these transitions there are pronounced oscillations in enzyme activity levels (characteristic for each enzyme monitored) which gradually decline in amplitude as the new steady state is established. How far this behaviour is to be explained in terms of changes in rates of enzyme synthesis and degradation has yet to be determined. Clearly study of these transitions will yield entirely new data on metabolic regulation in higher plant cells; whether it will yield the key to expose the changes underlying cytodifferentiation is less certain.

Turbidostat Cultures

The constancy of culture opacity in the steady state described has formed the basis for a second form of continuous culture—the turbidostat system (a continuous system in which inflow of new medium occurs in response to an increase in the opacity-decrease in the light transmission of the culture). The turbidostat culture system developed for work with sycamore cell cultures. Here, each time the population density exceeds a pre-selected value as determined by the optical density continuously monitored by the photocell, an electronically operated valve opens to admit a pulse of new medium. This reduces the optical density of the culture and the valve closes. These imputs of new medium are balanced by periodic harvesting of small volumes of culture in response to an electronically controlled level detector which controls the output valve.

The effect is to produce a culture of constant volume and constant population density growing at a constant rate. Whereas in the chemostat growth rate must always be below μ_{max} (the culture growth being limited by the supply of a chosen nutrient) here in the turbidostat growth can safely proceed under non-limiting nutrient conditions and one can study the effect of physical factors (e.g. temperature, light regime, CO_2 tension) and growth regulators, in particular plant growth hormones, on growth rate (King, 1976), and by appropriate techniques (Gould, Baylis & Street, 1974) on the duration of the different phases of the cell cycle. This is the system where one can attempt to achieve conditions under which biomass increase expresses the maximum genetic potential for cell growth. Such a system offers an entirely new approach to studies on the molecular basis of the hormonal control of cell growth.

SYNCHRONOUS CELL CULTURES—STUDY OF THE CELL CYCLE

As indicated earlier in this chapter, plant cell cultures are routinely propagated by batch cultures initiated at a relatively high initial cell density (2×10^5 cells ml^{-1}) and subcultured when they enter stationary phase. It is, however, possible to initiate such batch cultures of sycamore cells at much lower cell densities (the minimum inocular density being *ca.* $1.5–4.0 \times 10^4$ cells ml^{-1} in the standard medium; see Stuart & Street, 1969) and to use as inoculum cells maintained in stationary phase as long as possible without suffering decline in viability.

When this is done using an enriched synthetic medium (Stuart & Street, 1971) the cultures show a more extended lag phase and then embark upon a succession of highly synchronous divisions as evidenced by the data for cell counts, mitotic index (percentage cells in a recognizable stage of mitosis) determinations and estimations of nuclear DNA content (by microdensity following Feulgen staining). By using for such synchronous batch cultures the 4-litre culture vessel developed for the continuous culture systems but modified by adding a stainless-steel sampling valve automatically operated by a timing device (Wilson, King &

Figure: Growth curve of a synchronous cell culture of sycamore (A. pseudoplatanus) showing log cell number ml^{-1} with time. The interphase plateaux separating the synchronous increases in cell number are labelled P_1 –P_4. The lag phase of this culture was of 9 days duration and the culture reached stationary phase on day 24.

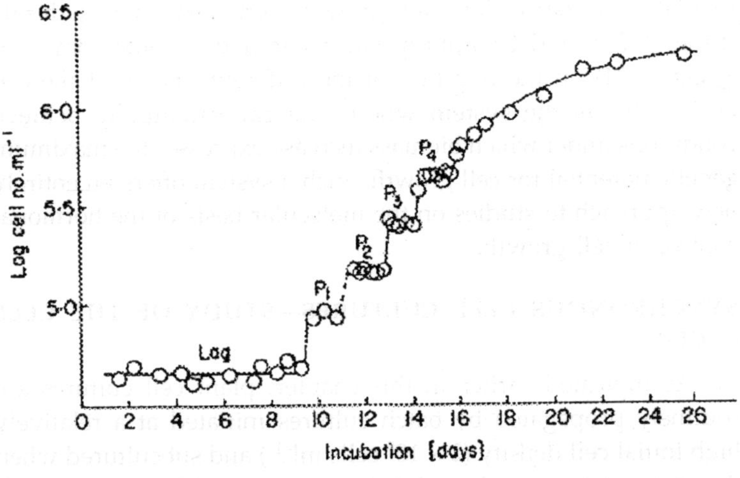

Street, 1971) it has been possible to show the synchrony of a number of metabolic events in the cell cycle (King *et al.*, 1974).

The period during which DNA is doubled takes place during a restricted period of the interphase between successive mitoses. Howard and Pelc (1953) termed this the S phase and distinguished the interphase period before S phase as G1 (the first gap) and the period after completion of S phase and before mitosis (M) as G2 (the second gap). Cell cleavage (cytokinesis) usually follows directly upon mitosis so that G1 can be timed from the origin of the daughter cells although, as originally defined, G1 begins as soon as nuclear division is complete.

These stages in the cell cycle can be determined in exponential asynchronous cultures by determinations of doubling time (cell count data), mitotic index, labelling index (per cent nuclei and mitoses labelled following flash label with tritiated thymidine-- ^{3} H-Tdr), fraction of labelled mitoses (following pulse-labelling with ^{3} H-Tdr; see Quastler & Sherman, 1959; and continuous labelling with ^{3} H-Tdr; see Cleaver, 1967), microdensitomitry and autoradiography, achieved by the single slide technique of Mak (1965).

Figure: (A) Mitotic index fluctuations for P_2 and P_3 of Fig. 15.6. (B) Average nuclear DNA contents over the same period. G1 and G2 values for nuclear DNA content of the cell line are shown for comparison. [C] represents the durations of the two successive periods of cell number increase. (Data from Gould & Street, 1975.)

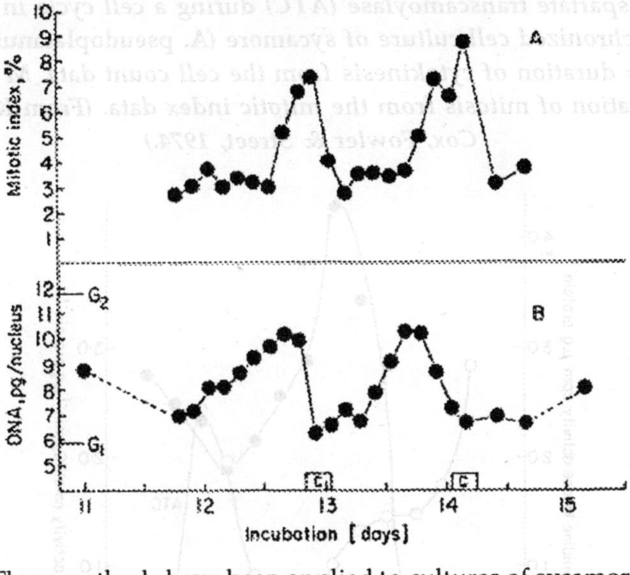

These methods have been applied to cultures of sycamore cell lines (Gould, Bayliss & Street, 1974) during the phase of exponential growth in batch cultures and to steady state chemostat cultures growing at different rates (doubling times ranging from 22 to 85 hr).

This has revealed that the phases S (7.0 ± 0.2 hr), G2 (8.7 ± 0.6 hr) and M (2.9 ± 0.3 hr) are relatively constant whereas, according to the cell doubling time, G1 varies widely (4–60 hr).

This observation that G1 varies with different cycle times, whereas S + G2 is relatively constant has also been observed in work with mammalian cells both in culture and *in vivo* (Mitchison, 1971). Certain critical events in G1 may be essential to the initiation of S phase and may be rate limiting.

In the work with sycamore cells growing at reduced rates, nitrogen supply is the limiting factor and when cells enter stationary phase in batch culture, they are arrested in G1 (this arrest is important in relation to the synchrony which can be achieved by regrowth of nitrogen-starved cells; see Gould & Street, 1975).

Figure: Changes in the activities of thymidine kinase (TK). and aspartate transcamoylase (ATC) during a cell cycle in a synchronized cell culture of sycamore (A. pseudoplatanus). C = duration of cytokinesis from the cell count data. M = duration of mitosis from the mitotic index data. (From King, Cox, Fowler & Street, 1974.)

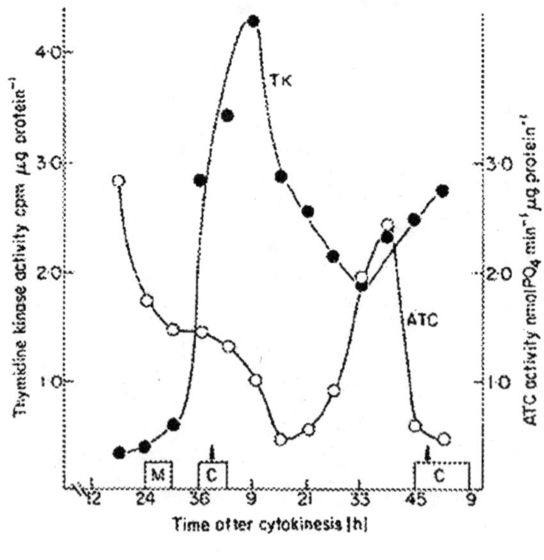

The value of synchronous cultures for studies on the cell cycle is that they enable particular metabolic events to be monitored. Work along these lines with plant cells has only recently been undertaken. In a study involving synchronous sycamore cell cultures (King *et al.*, 1974) it was shown that total extractable protein and RNA rise throughout interphase but at an increased rate during S + G2. This was supported by studies of the rate of incorporation of labelled amino acids and uridine into these fractions. Respiration rate similarly rose throughout interphase but with two peaks of activity during S phase and cytokinesis.

When, however, changes in the activity of extracted enzymes were studied different patterns emerged. Thymidine kinase and aspartate transcarbamoylase showed single and separated peaks of activity, succinic dehydrogenase showed two well separated peaks of activity, and glucose-6-phosphate dehydrogenase activity rose continuously throughout interphase.

Whilst the content of total extractable histone rose parallel with total protein, study of the rates of incorporation of ^{3}H-labelled lysine and ^{14}C-labelled arginine into this fraction pointed to greatly enhanced histone turn-over associated with S-phase (Street, Gould & King, 1975). It is not known whether the changes in enzyme activity detected in these studies reflected changes in rates of enzyme synthesis. Yeoman (1974) in work on cell synchrony in explants of Jerusalem artichoke tubers has shown however, by using the deuterium labelling technique (Hu, Bock & Halvorson, 1962), that changes in the activity of glucose-6-phosphate dehydrogenase during interphase do result from changes in rate of synthesis of the enzyme.

If future work along these lines enables the many separate metabolic events of the cell cycle of plant cells to be chartered, we will then have a number of 'markers' (points where particular metabolic events are initiated or terminated). This will enable us to identify those events along the interphase plateau whose initiation or pace is affected by nutritional factors and plant hormones, to determine whether certain events only occur when their controlling genes (which could be 'mapped') are exposed during DNA replication and whether other processes which proceed continuously show gene dosage effects as evidenced by increases in their rate after the points in DNA replication when their controlling genes are duplicated.

MORPHOGENESIS IN CELL CULTURES—CONCEPTS OF TOTIPOTENCY AND DETERMINATION

Various forms of morphogenesis have been observed in cell suspension cultures; this morphogenesis is apparently dependent upon the development in the cultures of relative large aggregates of cells, often several hundred in number, associated in a symplast. The morphogenesis can take the form of root initiation, shoot bud initiation or the development of somatic embryos (often referred to as embryoids) (Konar, Thomas & Street, 1972). Usually only one form of morphogenesis is expressed; sometimes the situation is more complex but in these cases one form is usually dominant—carrot cell cultures can be manipulated to show predominantly root initiation or exclusively embryoid development (Kessel & Carr, 1972).

Somatic Embryogenesis

Work on embryogenesis in carrot cultures illustrates clearly the problems in cell physiology raised by the morphogenetic potential of plant cell cultures. These studies date from the report in 1958 (Steward, Mapes & Mears, 1958; Steward, 1958) of embryo-like plantlets in liquid carrot cultures and their continuing development when the cultures were plated out on agar-solidified medium. Further work quickly established the presence in such cultures of structures strikingly similar to the globular, heart-shaped and torpedo-shaped stages of normal embryology from the zygote.

Halperin and coworkers (Halperin & Wetherell, 1964; Halperin, 1967; Halperin & Jensen, 1967) and Street and coworkers (Smith & Street, 1974; McWilliam, Smith & Street, 1974) have shown that the embryos in carrot cultures arise from single cells at the surface of the cellular aggregates and are released, at various stages of development, as free-floating structures which, if they have already reached the advanced globular stage, are capable of completing their development in isolation from the parent embryogenic clump. These cultures grow actively, the embryogenic clumps proliferating and fragmenting due to enlargement and separation of cells in their interior, and do *not* form embryos in a synthetic medium containing sucrose, inorganic salts, thiamine, *meso*-inositol, kinetin and auxin (2, 4-dichlorophenoxyacetic acid).

To initiate embryogenesis they are transferred to a similar medium, lacking auxin, which contains nitrate and ammonia (or urea or glutamine) and is at pH 5.0–5.4. Despite earlier claims, coconut milk is neither essential for growth nor for embryogenesis. Evidence that immature embryos have exacting nutritional requirements (isolated embryo culture—see Street, 1969) and the contrasted simplicity of the culture medium which supports prolific embryogenesis in carrot cultures suggests that the embryogenic cell aggregate fulfils a 'nurse' role to the embryogenic surface cells and that the embryos must remain attached to the aggregate to complete their early development to the point where they can survive and mature into plantlets when released.

This phenomenon of somatic embryogenesis raises a number of questions of cellular physiology. The term totipotency has been introduced to describe the embryogenic competence of the single cells from which the embryos arise. At present it is only possible to establish new plants, whether *via* embryogenesis of *via* shoot bud initiation, followed by adventitious root development, from the cell cultures of a limited (if now quite large) number of species or varieties within species.

Cultures which show this morphogenetic potential support the view that the pathways of cytodifferentiation which result in living tissue cells do not involve any loss or permanent inactivation of the genome and that they are, under appropriate environmental stimuli, completely reversible.

Whilst however some cell cultures remain recalcitrant, this cannot be established as a universal principle. It may be that in such cases the conditions of culture fail to provide or permit the synthesis of an essential morphogen. Halperin (1967) has, on the other hand, advanced the very interesting hypothesis that the achievement of totipotency occurs during the initiation of the carrot culture from the primary explant (storage root or seedling organ) and that embryogenesis is expressed by cell clumps derived from such 'induced' cells, the primary culture consisting of both these and 'non-induced' cells.

Retention of high embryogenic capacity in the cultures will then depend upon culture conditions favouring the active proliferation of the induced cells. On this hypothesis, failure to obtain embryogenesis in culture would have as its primary cause inappropriate conditions of callus initiation; the conditions of initiation would have effectively activated cell division and growth in the explant cells but failed to achieve the necessary 'dedifferentiation' to obtain cells with the competence of the zygote (a concept already raised here in previously discussing cytodifferentiation in cell cultures).

In classical plant embryology it has been considered that the early segmentations of the zygote (at least up to the 16-celled proembryo) follow a precise and species specific sequence which has phylogenetic significance (Johansen, 1950). During these divisions the cells of the proembryo are considered to inherit different cytoplasmic potentialities from the different regions of the zygote and these differences are regarded as determining from the beginning the exact role they and their daughter cells will play in constructing the embryo and its parts. This concept is often referred to as the theory of precise mosaic organization. The early segmentations involved in somatic embryogenesis have now been followed in a limited number of species (Street, 1976).

These segmentation sequences involved in somatic embryogenesis show more uniformity one with another than is depicted in the published accounts of the zygote embryology of the species concerned. In summary, early development of somatic embryos involves enlargement and regular segmentation in a subspherical cell mass by walls of minimum surface. This supports the concept of D'Arcy Thompson (1942), based upon his studies in animal embryology, that surface tensions are important in determining the early segmentations of embryology and that only at a later stage do localized growth centres emerge whose functioning gives rise to the divergences in the morphology and anatomy of embryos of different species.

This regulative theory of organization regards the early segmentations as being controlled by physical factors and as not involving any 'determination' of the early formed cells. With this background, the greater diversity and specificity recorded for zygotic embryogenesis can be interpreted as resulting from the physical restrictions and polar chemical gradients imposed upon the embryo as it develops within the ovule; when these influences are removed, as in cell cultures, the embryology reverts to a more basic or 'primitive' type of segmentation. This interpretation is supported by observations on the segmentations observed in natural polyembryony and by recent reports that indeed much more variable patterns of segmentation occur in ovule embryology than has hitherto been recognized (Jensen, 1965; Brown & Morgensen, 1972).

Polarity of Embryogenic Cells

The recognition that the embryos arising in cell culture have their origin in superficial cells of the cell aggregates raises the question of whether such cells have any unique cytological characters and whether their observation can yield information on the physical basis of cell polarity. In carrot cultures proliferating in the presence of 2, 4-D (Street & Withers, 1974) these cells are small, rich in cytoplasm and with a large diffusely-staining nucleus containing a prominent nucleolus. Small vacuoles are clustered round the nucleus and each cell contains several amyloplasts containing prominent starch grains. Study of these cells in the electrorn microscope shows the presence of numerous round and oval mitochondria and Golgi bodies and the regular presence of small numbers of lipid bodies (spherosomes). As might be expected from their meristematic activity, these cells are frequently observed in mitosis (in contrast to the expanded interior cells of the aggregates) and show numerous wall microtubules and limited arrays of cytoplasmic and nuclear microfibrils. When the cultures are transferred to auxin-free medium there is a transient increase in proliferation prior to the initiation of embryogenesis and the superficial cells show a change in segmentation pattern leading to the origin of 4-celled groups.

The individual cells in these groups either initiate an embryo, or by their further division promote the growth of the aggregate or undergo expansion (and senescence?) and become involved either in the release of the developing embryo or the break-up of the proliferating embryogenic aggregate. Associated with this changed segmentation pattern the densely cytoplasmic superficial cells show changes in fine structure. They have increased numbers of E.R. profiles (parallel arrays of rough E.R. profiles become particularly prominent) and of ribosomes. The Golgi bodies are more numerous and more compact. Additional large mitochondria (discs with swollen rims in outline) make their appearance and may come to occupy a considerable volume of the cytoplasm. The cells of the young proembryos show similar fine structure.

Although the first division of the embryogenic cells is by a wall parallel to the surface of the aggregate and at right angles to the longer axis of the cell (giving rise to an apical and a basal cell; the former being the first cell of the proembryo proper and the latter of the very variable suspensor), nevertheless fine structure studies do not reveal any prior asymmetry in the distribution of cytoplasm or cell organelles. Rather disappointingly, these studies have not revealed any unique features of the embryogenic cells or exposed any structural polarity. Perhaps this is not unexpected when we bear in mind the lack of any uniformity of fine structure in those angiosperm zygotes which have been studied (Jensen, 1965; Schulze & Jensen, 1969; van Went, 1970; Morgensen, 1972). Such studies only serve to emphasize that the special nature of embryogenic cells must now be approached at the level of molecular biology.

POLLEN GRAINS AS ISOLATED EMBRYOGENIC CELLS AND AS A SOURCE OF HAPLOID CELL LINES FOR MUTAGENESIS

This field of study was opened up by the pioneering studies on anther culture by Guha and Maheshwari (1964) working with

Datura, and Bourghin & Nitsch (1967) working with *Nicotiana.* They showed that anthers, excised at an appropriate stage in pollen grain development and cultured in a simple medium gave rise to haploid embryos derived from individual pollen grains. Haploid plantlets can now be obtained by this technique from many species within the family Solanaceae. This approach extended to species in other angiosperm families has in a few cases yielded haploid callus but more frequently given a negative result.

If we assume that immature pollen grains can, under appropriate stimuli, embark upon embryogenesis (sporophyte development) then the difficulty of achieving this with most excised anthers may be because conditions within the anther are too strongly promotive of the gametophyte pathway of development (i.e. that which gives functional pollen grains containing a tube nucleus and a generative cell).

Recently however Nitsch (1974) in work with two Solanaceous species *Nicotiana tabacum* and *N. sylvestris* has shown that if anthers are excised at the peak of the first pollen mitosis, kept at 5°C for 48 hours and cultured for 4 days in a simple medium, then the immature pollen grains can be extracted. If this pollen grain suspension is cultured in a thin layer of liquid medium containing IAA, zeatin, glutamine and serine as supplements, some 10% of the grains embark upon embryogenesis. Although this work has so far been restricted to Solanaceous plants, it could prove to be the key to inducing embryogenesis or haploid callus initiation in pollen grains of the many species resistant to the culture of whole anthers.

The importance of these studies is twofold. First is their potential for providing haploid plants (and by appropriate treatment homozygous diploids) and haploid cell cultures from many species and varieties. The haploid cell lines could be used as a source of mutant cell lines extending dramatically the field of the biochemical genetics of higher plants. Such mutant haploid lines could be rendered diploid by colchicine treatment and hence used to generate fertile mutants, provided their potential for morphogenesis was not impaired by the mutation.

They could be preserved as cell culture lines by freezing preservation (Nag & Street, 1973, 1975a,b) as is already routinely done with mutant lines of bacteria. The use of such mutant cell lines in the continuous and synchronous culture systems previously described would permit more critical studies to be undertaken on particular metabolic sequences and their control and on the biochemical nature of the controls which operate in cell growth and division. Work along these lines is already actively proceeding (Widholm, 1974; Street, 1975b) but is currently handicapped by the need for more effective means of selecting the desired mutant cells which are present in very small numbers in the cell populations after mutagen treatment (e.g. 1 in 10^7).

The second important feature of these studies is in relation to embryogenesis. If within the pollen suspension, the pollen grains destined to embark upon embryogenesis are of characteristic size and/or density, it may be possible to isolate them by appropriate density gradient centrifugation. This would enable us to characterize embryogenic cells by the techniques of molecular biology, and to study in greater detail the segmentation patterns of early ermbryogenesis and the spontaneous expression of polarity in the proembryonal cell mass. Since the early stages of embryogenesis in pollen occurs within the enclosing pollen grain wall, study of the food reserves of such grains and of their biosynthetic activity may enable us to determine the special requirements of the proembryo.

This chapter has drawn attention to the availability of new experimental material and of new techniques in the field of plant cell culture. The exploitation of this approach in the molecular physiology of plants is still in its infancy. It is therefore important that the new generation of plant physiologists and biochemists should remain well informed of its present transient limitations and assured future prospects.

6

ISOLATED PLANT PROTOPLASTS AND PHYSIOLOGY

INTRODUCTION

The isolated plant protoplast is a single cell, bounded by the plasmalemma and, when first formed, containing all the normal cell components, although being separated completely from its cell wall. From a physiological viewpoint, however, the protoplast cannot be regarded simply as a cell lacking a wall, since the mechanics of isolation, in conjunction with environmental factors, undoubtedly influence its metabolism and elicit subtle ultrastructural changes. The absence of a functional cell wall may affect the permeability of the cell membrane and lead to a general leakage of solutes from the protoplast. The protoplast is also in a transient state since most protoplasts, irrespective of their immediate cultural environment, will initiate the synthesis of a new cell wall a few hours after release, and eventually revert to a single-walled cell. In spite of these considerations, and combined with a cautious extrapolation of experimental data to the presumed situation existing in the intact cell or whole plant, protoplasts provide an important biochemical tool for the biologist.

In the absence of a cell wall, the exposed plasmalemma can be examined in great detail with respect to particle uptake, permeability, possible membrane-associated functions such as disease resistance, membrane fusion, and, during cell wall synthesis, the relationship of the plasmalemma to its cell wall.

ISOLATION AND CULTURE OF PROTOPLASTS

Preparation of Protoplasts

Protoplasts may be produced, under aseptic conditions, from a wide range of plant species either directly from the whole plant, or indirectly from *in vitro* cultured tissues. There are two basic approaches for the enzymatic isolation of protoplasts: (a) the treatment of a plant tissue with a mixture of pectinase and cellulase so as simultaneously to macerate, or separate, cells and degrade their walls (Power & Cocking, 1970); and (b) the sequential (two-step) method involving the production of isolated cells which in turn are converted into protoplasts by a cellulase treatment (Nagata & Takebe, 1970).

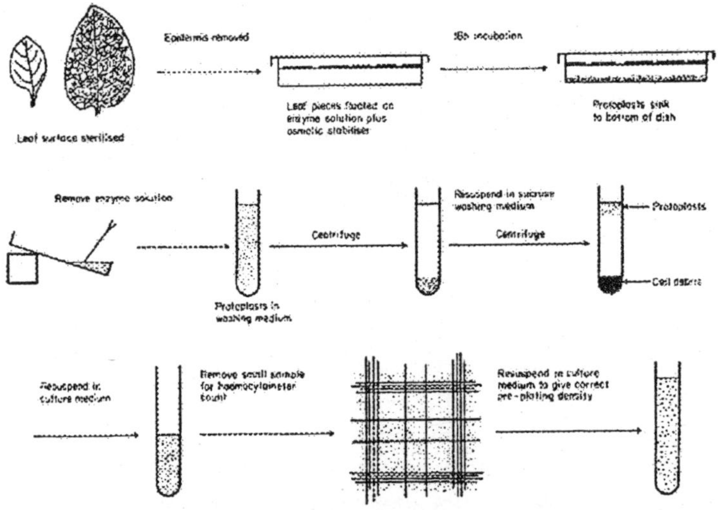

Figure: Generally applicable scheme for the isolation, washing and counting of leaf mesophyll protoplasts. Protoplasts isolated from callus or cell suspensions are handled in the same way.

Since removal of the cell wall results in loss of wall pressure upon the cell, protoplasts are isolated and maintained in hypertonic plasmolytica provided by a balanced inorganic salt medium or monosaccharide sugar solution. Mannitol, for example, is not readily transported across the plasmalemma and therefore provides a stable osmotic environment for the protoplast.

The enzymes used for the isolation of protoplasts are crude extracts of fungal origin. The pectinases are rich in polygalacturonidase activity whilst the cellulases contain hemicellulase, *b* -1,4-glucanase, chitinase, lipase, nucleases and pectinase.

A generally applicable scheme for protoplast production, using the mixed enzyme procedure. In order for the enzymes to gain access to the plant tissues, as in the case of leaf palisade and mesophyll cells, the lower epidermis must be removed by peeling or partial digestion with cutinase. Certain types of leaves, particularly of the cereals, must be sliced prior to enzyme incubation, since the epidermis cannot readily be removed. Most plant tissues and organs (roots, root nodules, coleoptile, leaf epidermis, petals, germinating pollen grains, fruit placenta, tetrads and microsporocytes) will yield protoplasts after suitable adjustment of the enzyme mixture and plasmolyticum. For example, the pollen tetrad wall consists of callose and so only an enzyme rich in *b* -1, 3-glucanase (snail digestive juice enzyme) will liberate protoplasts.

Protoplasts are produced from calluses and cell suspensions (Wallin & Eriksson, 1973) often only during the log phase in their growth cycles, since the composition of the primary cell wall varies as secondary products, such as lignin, are deposited as the culture matures, rendering it unsusceptible to complete degradation by cellulase. In general, the production of protoplasts from an untried source will always involve a consideration of enzyme purity, pectinase to cellulase ratios, protoplast yield and viability.

Following enzyme incubation, the spherical protoplasts must be separated from cellular debris, subprotoplasts and vacuoles. Subprotoplasts are formed during early plasmolysis when the protoplast splits into two or more subunits, some of which will be enucleate and hence non-viable. Separation can readily be achieved in a variety of ways: repeated resuspension and centrifugation of protoplasts in a washing medium; flotation of protoplasts on a hypertonic sucrose solution; passage of the incubation mixture through a nylon sieve of suitable pore size which allows only protoplasts to pass through; or by the use of a two-phase system, such as dextran-PEG (polyethylene glycol), which is based upon a density difference between protoplasts and cells (Kanai & Edwards, 1973).

Freeze etched protoplasts, collected after washing and examined in the electron microscope or treated with fluorescent brighteners which specifically bind to cellulose, reveal the complete absence of cellulose fibres on the plasmalemma surface.

Cell Wall and Whole Plant Regeneration

Protoplasts are cultured in either liquid or agar solidified nutrient media. The culture media are very similar in composition to those required for the *in vitro* culture of cells, but with the addition of an osmotic stabilizer to prevent bursting. During culture, up to 90% of protoplasts isolated from highly differentiated cells undergo a rapid and immediate process of dedifferentiation whilst at the same time initiating the synthesis of a new cell wall.

Early stages of wall synthesis are preceded by extensive infoldings of the plasmalemma together with an accumulation of pectin-like substances in vesicles found in the peripheral layer of cytoplasm. These early stages of wall synthesis, detected after 18 hours, are unaffected by the presence, in the culture medium, of protein synthesis inhibitors, suggesting that synthesis of new RNA or protein is not required for wall initiation and that residual protein and endogenous hormone levels are sufficient. Structurally, the first formed envelope is amorphous and consists of pectins, but after a few days a second inner layer of cellulose fibrils is progressively laid down on the protoplast surface, eventually producing a near normal cellulose matrix after four or five days (Burgess & Fleming, 1974).

Nuclear division and cytokinesis is concomitant with cell wall formation. Occasionally cytokinesis does not occur, perhaps due to the presence of an incomplete cell wall, and gives rise to binucleate cells which may not be capable of further division.

Most protoplasts, like suspension cultured cells, have an optimum plating density as regards their division potential. At this density (usually $5 \times 10^4 - 1 \times 10^5$ protoplasts/ml $^{-1}$), the plating efficiency, defined as a percentage of the original plated protoplasts that produce cell colonies after 28 days culture, can reach 70%. As division proceeds, the plasmolyticum level in the culture medium is reduced stepwise by the transfer of agar blocks, containing the dividing protoplasts, to the surface of a similar medium of lower osmotic pressure. Individual cell colonies are pricked out, 4–6 weeks after plating, and subcultured onto a regeneration medium. Callus masses thus produced may undergo differentiation, via organogenesis, eventually producing plantlets. Regenerating protoplasts of certain species, such as carrot, will form embryoids in liquid culture.

Whole plant regeneration from protoplasts can be achieved for an expanding list of species of distinct taxonomic groups including the Solanaceae (tobacco, *Petunia*) (Frearson *et al.*, 1973), Brassicaeeae (rape) (Kartha *et al.*, 1974), Papilionaceae (pea, cowpea) and Scrophulariaceae (*Antirrhinum*). Whole plant regeneration is not restricted to monocotyledons or dicotyledons, haploid or diploid plants, or to protoplasts obtained directly from whole plant tissues. The majority of plants regenerated from protoplasts are normal and exhibit a high degree of fertility, but a small and variable (less than 0.1%) number of plants show morphological abnormalities owing to aneuploidy and polyploidy.

An example of this can be seen following the regeneration of plants from mesophyll protoplasts of diploid *Petunia*, homozygous for blue flower colour. Some regenerated plants have a tetraploid chromosome number, which is also accompanied by a flower colour change from blue to red. Since well established protoplast cultures have high plating efficiencies, and hence efficient cloning potentials, abnormal plants can arise from the expression of cell aberrations present in the tissue prior to protoplast isolation.

The nutritional and hormonal requirements of cultured protoplasts are constantly varying depending upon the stage in the regeneration process. The photosynthetic capacity and respiration rate of the protoplast is suppressed by the plasmolysing conditions. Changes in metabolic requirements during the early stages of growth could be attributed to enzyme uptake during isolation and the rigours of cell wall synthesis. These subtle changes are highlighted in the crown gall (tumour) (Scowcroft *et al.*, 1973), where protoplasts require an exogenous growth regulator supply for the initiation of division, unlike the tumour cells which are autotrophic for auxin. Following cell wall regeneration and division, the requirement for exogenous growth regulator supply is lost, suggesting that growth regulator autonomy is, at least in part, dependent upon the presence of a functional cell wall.

PROTOPLASTS AND AUXIN RESPONSES

Although protoplasts require both auxins and cytokinins for the initiation and maintenance of division following cell wall formation, the freshly isolated protoplast has facilitated studies of the direct and rapid physiological action of auxins upon the plasmalemma. Auxins such as IAA are known to influence cell elongation in the plant. Leaf or coleoptile protoplasts, incubated in the presence of physiological levels of IAA, respond by an increased permeability of the plasmalemma, probably to water.

Protoplasts maintained in an isotonic plasmolyticum containing auxin, undergo a rapid expansion as the internal volume increases owing to extensive vacuolation, which often results in bursting (Hall & Cocking, 1974). The presence of anti-auxins, which preferentially bind to the site of action of the auxin, suppresses the bursting response indicating that these effects are the result of auxin action. As may be expected, protoplasts isolated from tumour cells which have high endogenous levels of auxins, require significantly higher levels of exogenously supplied auxin to elicit the bursting response. Vacuoles produced by disruption of protoplasts show no response to auxins, indicating clear differences in structure and function of the tonoplast. It is difficult to relate these responses to the auxin control of growth in the plant, but the induction of membrane convolutions, seen prior to bursting, can influence the thickness of the newly synthesized wall and hence cell shape and direction of growth.

Herbicides, such as 2,4-D, are used as growth regulators at low concentrations in culture media. There appears therefore to be a balance, for some species, between controlled growth and a herbicidal effect. Protoplasts, when incubated in the presence of herbicides such as paraquat, do not respond by expansion, but collapse as a result of disruption of the plasmalemma. It may be therefore, that control of the selective action of some herbicides is found in the plasmalemma itself.

UPTAKE PROPERTIES OF PROTOPLASTS

Macromolecules

During isolation, and for several hours after release, protoplasts exhibit extensive pinocytic activity. This process, which may be stimulated by the plasmolysing conditions, enables macromolecules, such as ferritin and polystyrene latex particles, to enter the protoplast. Infoldings of the plasmalemma result in the formation of vesicles which contain small quantities of the surrounding medium together with any particles that may be suspended in the medium and which are less than I μm in size. Particles entering protoplasts in this way are often accumulated in the central vacuole. These basic studies have led on to an examination of uptake in relation to cell and ultimately whole plant modification, by the incorporation into protoplasts of biologically active particles. Certain microorganisms can be transformed by the incorporation into the genome of exogenously supplied DNA, which in turn can be expressed, thus modifying the cell. Comparable attempts at transforming eukaryotic cells, involving plant protoplasts, are in progress and it is clear that DNA will enter protoplasts but no evidence for the persistent expression of this input DNA currently exists at the protoplast/cell level.

Chloroplasts, Nuclei and Bacteria

A further extension of the uptake phenomenon has been successfully applied to whole bacteria and plant organelles such as nuclei and chloroplasts (Bonnett & Eriksson, 1974). The precise mechanism of entry of these relatively large particles may be one of membrane fusion, in the case of nuclei and chloroplasts, rather than pinocytosis. Conditions favouring organelle uptake are very similar to those that bring about protoplast fusion. Nitrogen fixing bacteria (*Rhizobium*) can enter pea leaf protoplasts during the isolation procedure (Davey & Cocking, 1972) but evidence for nitrogen fixing ability of the bacterium/plant association is still awaited. Likewise transplanted nuclei may survive degradation, and following fusion with the nucleus of the recipient cell, can result in the formation of hybrid or modified cell lines. Organelles entering protoplasts as a result of uptake into vesicles may not be able to function in a strictly symbiotic manner since they are not in direct contact with the cytoplasm. However, these novel approaches to cell modification may have major implications for the plant breeder and could possibly lead to the production of more efficient economically important plant species.

Virus Particles

The availability of protoplasts in large quantities has overcome many biological problems facing the investigator of host-virus relationships. In intact plants the process of infection is not synchronous and its establishment is restricted to cells with broken or damaged cell walls (Zaitlin & Beachy, 1974). Protoplasts become infected following a short incubation with intact virus or its RNA and in the presence of a polycation, such as poly-l-ornithine, which not only damages the plasmalemma but neutralizes the negative charge common to many virus particles. This allows adsorption of particles, to the plasmalemma surface, which can then enter the protoplast through damaged regions of the membrane or *via* pinocytic vesicles. The process of infection varies considerably depending upon the virus, but in general, newly synthesized material, possibly RNA, is detected after 2–3 days in vesicles closely associated with the endoplasmic reticulum and nucleus. Newly formed virus particles are also found within the nuclear membrane. However, it is impossible to eliminate influences of the plasmolyticum, enzyme carryover and the absence of a cell wall upon the infection process, which may therefore be dissimilar to that in whole plants. Some protoplasts resist infection with virus, and it may be possible to regenerate plants, as described previously, from such protoplasts, which are not susceptible to virus infection.

SOMATIC HYBRIDIZATION OF PLANTS

Crop improvement advances with the production of new hybrids, which, in turn, are selected for the expression of desired characteristics. However, in order that plant species can preserve their genotypes there exists major incompatibility barriers. The fusion of plant protoplasts of different species offers one approach that could by-pass the sexual cycle and, provided that nuclear fusion occurs in the resultant heterokaryon, new somatic hybrids may arise following regeneration.

Induced Fusion of Protoplasts

Several methods now exist for the fusion of protoplasts and all embrace the same basic concepts. It is first necessary to bring the protoplasts together so that the two plasmalemma surfaces are in contact, whilst at the same time inducing destabilized regions on the membrane. Where two such regions of adjacent protoplasts coincide, localized membrane fusion occurs, together with limited cytoplasm mixing, to produce cytoplasmic bridges. Further mixing of the cytoplasms eventually results in the complete coalescence of the protoplasts which are then returned to stabilizing conditions after removal of the fusion inducing agent.

A slightly hypotonic aqueous sodium nitrate solution reduces the negative charge on the outer membrane thus allowing adhesion of protoplasts. The presence of the Na^+ ion induces localized realignment of the lipo-protein in the membrane which in turn allows fusion to occur when two such regions come into contact. Slightly deplasmolysing conditions facilitate cytoplasmic coalescence (Power & Cocking, 1971).

PEG solutions induce tight adhesion of protoplasts (Kao *et al.*, 1974) owing to electrostatic forces. The presence of cations, such as Ca^{2+}, in the PEG solution, enhances adhesion since the calcium ion may form a molecular bridge between the negatively polarized PEG molecule and the protein of the membrane. Water is also withdrawn from the protoplast by PEG, thus reducing the turgidity of the system and allowing closer packing of protoplasts. This dramatic disruption, and possible reversal, of the charge properties of the membrane after removal of PEG, leads to membrane fusion.

Incubation of protoplasts in solutions buffered at high pH (Keller & Melchers, 1973) and containing calcium ions also induces fusion. This may resemble the action of PEG in reversing membrane charge properties in conjunction with the establishment of molecular bridges prior to membrane fusion.

Protoplasts isolated from pollen mother cells will fuse in the absence of an inducer. In such cases the membrane structure may be quite different to the plasmalemma of somatic cells since the products of pollen mother cell development, germinating pollen grains, are ultimately involved in the fertilization process and hence membrane fusion.

Culture of Fusion Products

Following fusion treatment, protoplasts are returned to the culture medium and cell wall formation proceeds. It is during the first mitotic division that nuclear fusion can occur as a result of common spindle formation, giving rise to the hybrid cell. Abortion of the nuclei, unidirectional chromosome elimination, or the inability to produce a nuclear hybrid may be expected between species that are not closely related genetically. Following regeneration into whole plants, the resulting cytoplasmic hybrids ('cybrids') may be of immense value to the plant breeder since many plastid characteristics and resistance to certain diseases are controlled by plasmagenes.

The frequency of true hybrid cell formation is low (1 in 10^5). This fact has led to the development of generally applicable cultural procedures aimed at preferentially selecting out hybrid cells. These techniques, currently receiving most attention by protoplast workers, are based upon an ability of the nuclei to complement in the hybrid cell. This involves the use of parental species that exhibit mutually exclusive biochemical features such as differential drug sensitivity (Cocking *et al.*, 1974), light sensitivity or auxotrophic requirements.

Following selection, somatic hybrid plants have been produced between two *Nicotiana* species. In this particular case, the selection theory was based upon prior knowledge of the cultural requirements of the sexually produced hybrid (Carlson *et al.*, 1972). Protoplasts of the parents, *N. glauca, N. langsdorffi,* did not grow in a culture medium that supported the growth of leaf protoplasts of the sexual hybrid. Following fusion with sodium nitrate, the protoplasts were plated in this medium and a few colonies were recovered which later developed into plants. Electrophoretic and karyotypic analysis confirmed that these plants were amphidiploid somatic hybrids.

In spite of this unambiguous proof that somatic hybridization is possible, it will only be when somatic hybridization is demonstrated for plant species which are truly sexually incompatible, that its potential for crop improvement will be fully realized.

GENETIC VARIATION IN CULTURED PLANT CELLS

Genetic variability can be described as the quantity of alleles which exist among the total genetic loci of a population. In many crop plants genetic variability is rapidly diminishing (Day, 1972) as a result of increasingly specialized breeding, intensive agriculture, and wide dissemination of the most desirable cultivars. Serious consequences are uniform susceptibility to new pathogens, inability to adapt to new environments as well as the inadvertent spread of 'silent' deleterious genes. The southern corn blight, which destroyed one billion dollars worth of the U.S. corn crop in 1970, occurred in large part because 70% of the cultivated corn contained the same cytoplasmic determinant for male sterility. A pleiotropic effect of this gene is blight susceptibility.

While efforts are underway to preserve existing variability (e.g. in germ plasm collections) other approaches such as mutation breeding seek to increase variability. To this end plant cell culture is potentially well-suited. The plant cell geneticist can easily select certain mutants from among 10^4 –10^5 cells per ml of suspension culture. Each cell is a potential plant which, if field tested by conventional methods, would require huge investments of time, labour and acreage. The individual cell has limited metabolic pools and can be cultured on a completely defined and relatively simple medium.

The plant cell geneticist thus can establish a wide range of conditions to select for specific biochemical mutants which are perhaps not so easily selected in the field. The ability to select defined genetic alterations will provide a means for the dissection of biochemical and developmental pathways and for the analysis of mechanisms of genetic expression in higher plants. This knowledge will lead to efficient selection *in vitro* of mutations which result in improved nitrate utilization, higher heavy metal tolerance, over production of a specific amino acid or seed protein, and other valuable traits (Chaleff & Carlson, 1974). Theoretically, the use of tissue cultures from anther-derived haploid plants would allow the direct selection of recessive mutations. Finally, whole plant regeneration from callus is becoming possible in an ever-increasing number of species. This is essential if the microbial geneticist's approach in plant cell culture is to complement conventional plant breeding.

The following section discusses some mutants already obtained in culture and some advances in plant cell genetics that are important for producing additional agriculturally useful variation.

INDUCED VARIATION AND SELECTION

The spectrum of mutants recovered among cultured plant cells has so far been determined primarily by the growth characteristics and requirements of these cells *in vitro*. The strict prerequisite of a high minimum density for proliferation of cells (Street, 1973) or protoplasts (Nagata & Takebe, 1971) *in vitro* makes easier the selection for mutants resistant to antibiotics and antimetabolites than the selection for auxotrophs. The isolation of auxotrophs requires an experimental system composed of haploid cells in which the wild type cells are at a disadvantage compared to the mutants. This approach was used by Carlson (1970) to isolate six auxotrophic mutants of *Nicotiana tabacum*. The selective system was adapted from a procedure used successfully with mammalian cells (Puck & Kao, 1967). In a minimal medium containing 5-bromodeoxyuridine (BUdR), actively dividing prototrophic cells will incorporate more of this lightsensitive nucleoside analogue into DNA than will arrested auxotrophs.

The auxotrophs will survive subsequent illumination at a much higher frequency than the prototrophs and will form clones on a supplemented medium. The mutant tobacco calluses which were isolated grew slowly on a minimal medium. Plants were differentiated from four of the mutant clones and these also grew slowly without supplementation. In genetic analyses of these four regenerated plants, three plants transmitted the mutant phenotype as a single recessive Mendelian factor and one displayed a more complex pattern of inheritance. The recovery of only leaky mutants may be due to a lack of functional diploidization of the *N. tabacum* genome. Haploid cells obtained from this amphiploid species may contain two copies of many metabolically essential genes and thus two mutational events may be required to delete completely certain functions (Carlson, 1970).

Cell lines which exhibit increased resistance to a wide variety of antimetabolites have been isolated and studied. Haploid cultures of petunia (Binding *et al.*, 1970; Binding, 1972) and tobacco (Maliga *et al.*, 1973) resistant to streptomycin have been selected. In genetic crosses involving a diploid plant obtained from one mutant tobacco clone, drug resistance segregated as a maternally inherited character (Maliga *et al.*, 1973; Sz.-Breznovits *et al.*, 1974).

Plant cell lines containing dominant or semidominant genetic markers may prove useful in selecting hybrid cells formed by protoplast fusion. This realization has motivated the isolation of cell lines resistant to 8-azaguanine and BUdR since such mutants are employed in selecting hybrids in mammalian cell systems. Resistance to these analogue precursors of nucleic acids in cultured mammalian cells may result from deficiencies in hypoxanthine:guanosine phosphoribosyl transferase (HGPRT) and thymidine kinase activities, respectively, or in the uptake of these analogues (Kit *et al.*, 1963; Littlefield, 1964a; Breslow & Goldsby, 1969). In the presence of aminopterin, endogenous biosynthesis of thymidine and hypoxanthine is inhibited and cultured mammalian cells are dependent upon an exogenous supply of these substances.

As each mutant cell line is unable to utilize one of these compounds, only complementing hybrids formed by cell fusion survive in a medium containing hypoxanthine, aminopterin, and thymidine (the HAT system) (Littlefield, 1964b).

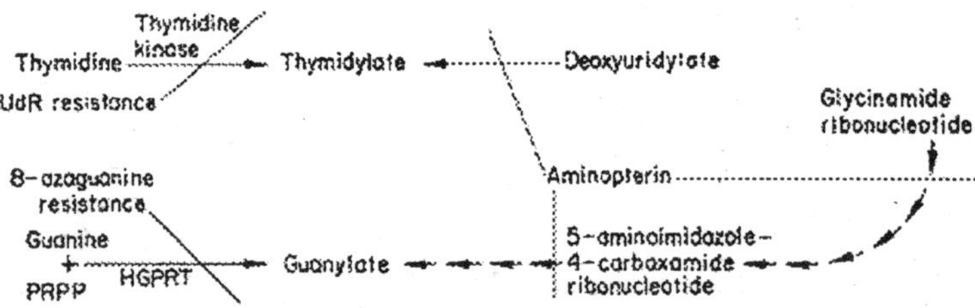

Figure: Simplified scheme showing pathways for synthesis of guanylate and thymidylate. Endogenous biosynthetic steps blocked by aminopterin-inhibition of tetrahydrofolate synthesis are shown as well as the genetic blocks which confer resistance to BUdR and 8-azaguanine.

Although a low concentration of aminopterin inhibits growth of sycamore cells (Bright & Northcote, 1974) it should not necessarily be assumed that this effect is caused by the same mechanism in both animal and plant systems. The isolation of cell lines of tobacco (Lescure, 1973) and sycamore (Bright & Northcote, 1975) resistant to 8-azaguanine has been reported. The resistance phenotype is not completely stable in the sycamore cell line in which HGPRT activi ty is half that of the wild type level. The latter result is not unexpected in a diploid system and emphasizes the necessity of using haploid cell lines in order to recover mutants with an anticipated recessive phenotype.

Resistance to BUdR has been selected in haploid tobacco (Maliga *et al.*, 1973) and in presumably diploid sycamore (Bright & Northcote, 1974) and soybean (Ohyama, 1974) cells. The resistant sycamore and soybean cultures did not differ significantly from the wild type in uptake of thymidine or in the level of thymidine kinase activity (Bright & Northcote, 1974; Ohyama, 1974) and the basis for the resistance phenotype remains unknown. Marton and Maliga (1974) have observed transmission of BUdR resistance to progeny of plants derived from one mutant tobacco callus and unimpaired uptake and incorporation of thymidine in four resistant cell lines. Heimer and Filner (1970) have isolated a cell line of tobacco which is insensitive to inhibition by threonine when grown in a medium containing nitrate as sole nitrogen source. By studying nitrate accumulation in this mutant under conditions in which nitrate reductase is chemically inactivated and in which nitrate uptake by wild type cells is repressed, it was concluded that regulation of nitrate uptake is altered in the resistant cell line. Such mutants should prove extremely useful in the investigation of nitrogen metabolism.

The potential agronomic importance of tissue culture techniques is illustrated by experiments in which screening of cultured somatic cells resulted in the recovery of mutants of *N. tabacum* resistant to wildfire disease (Carlson, 1973a). This disease of tobacco is caused by a bacterial pathogen, *Pseudomonas tabaci,* which produces a toxin structurally related to methionine (Stewart, 1971). Populations of mutagenized haploid cells were plated in a medium containing an inhibitory concentration of methionine sulphoximine, an analogue of the wildfire toxin (Braun, 1955) and resistant clones were selected. Three homozygous diploid plants which were regenerated from these methionine sulphoximine-resistant calluses are less susceptible than the parent plant to the pathogenic effects of bacterial infection. One mutant plant contains wild type intracellular levels of free methionine and resistance appears to be genetically complex. In the other two resistant plants the endogenous concentration of free methionine is five times higher than in the wild type while the free pool sizes of other amino acids remain unchanged (Carlson, 1973a). The levels of free methionine in the heterozygous plants are intermediate between those in the wild type and in the homozygous mutant (Chaleff & Carlson, 1975a).

Intermediate methionine levels are consistent with the observed transmission of the resistance phenotype of these two plants as a single semidominant locus (Carlson, 1973a). These experiments suggest that selection for toxin resistance *in vitro* may provide a generalized procedure for generating disease resistant varieties. It is also evident that regulation of amino acid metabolism may be altered in cells selected for resistance to an amino acid analogue and that this mutational event may be expressed stably in mature plants derived from the mutant callus tissue. Additional support for this conclusion is provided by other research.

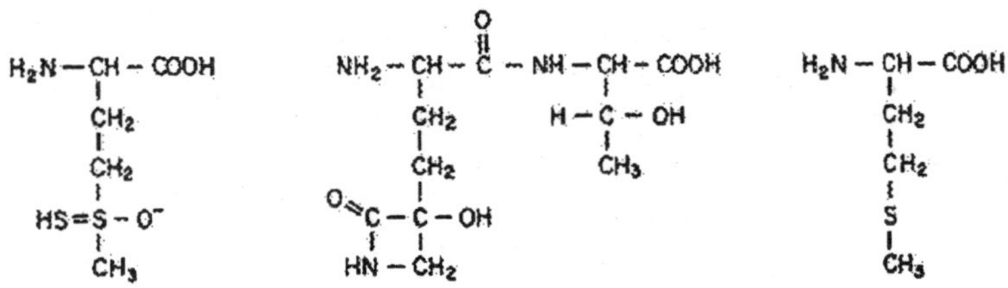

Methionine sulphoximine — Wildfire toxin (tabtoxin) — Methionine

Figure: Structures of methionine sulphoximine, wildfire toxin (Stewart, 1971), and methionine.

Widholm (1972a,b) has isolated presumably diploid cell lines of *N. tabacum* and *Daucus carota* which are capable of growth in the presence of a normally inhibitory concentration of 5-methyltryptophan. Crude extracts of the resistant cell lines (5-*mt* ^{r}) contain a species of anthranilate synthetase which is less sensitive to feedback inhibition by tryptophan and 5-methyltryptophan than is the wild type enzyme. Endogenous levels of free tryptophan in 5-*mt* ^{r} cell lines of tobacco and carrot are 15 times and 27 times higher, respectively, than the wild type levels. Plants could not be regenerated from the tobacco and carrot cells which synthesized a feedback-insensitive form of anthranilate synthetase.

However, plants were obtained from analogue-resistant carrot cultures in which anthranilate synthetase activity was indistinguishable from the wild type and uptake of 5-methyltryptophan and tryptophan was reduced. Although no genetic studies with these plants have been reported, cell lines re-initiated from the plants maintained the same degree of resistance to 5-methyltryptophan as the original selected lines. Cell cultures of tobacco and carrot resistant to *p* -fluorophenylalanine were isolated also. Resistance to this analogue apparently resulted from diminished rates of transport and incorporation (Widholm, 1974).

Two attempts have been made to employ analogue resistance as a selective screen in obtaining callus cultures of cereals which overproduce amino acids essential to human nutrition. Chaleff and Carlson (1975b) recovered three cell lines of rice which were stably resistant to the lysine analogue S-2-aminoethylcysteine. Analyses of both free pool and total amino acid compositions showed that the resistant cultures contained elevated levels of several amino acids, including lysine, methionine, leucine and isoleucine.

Cheng (1975) has selected cultures of *Hordeum vulgare* which are resistant to the same analogue and in which lysine levels were increased, as was the rate of incorporation of exogenously supplied lysine. Unfortunately, attempts to differentiate plantlets from the variant calluses were unsuccessful in both cases. Although much valuable information is to be obtained from research with variant cell lines, such studies are severely limited by the inability to examine genetic expression and transmission in the mature plant. Future investigations should attempt to utilize experimental systems in which plants may be differentiated from the altered cell lines.

GENE TRANSFER

The systems so far described have dealt with mutationally-induced variation of the pre-existing plant genome. Considering such factors as genome size, multiple gene copies (e.g. non-allelic isozymes), and the tendency of plants to become aneuploid and polyploid it should be apparent that the potential for induced variation in plant cells is enormous. As culturing and selective techniques improve and a better understanding of plant biochemistry is attained (largely *via* tissue culture) the spectrum of plant cell mutants recovered should widen. In addition, manipulation of plant cells offers unique opportunities to introduce foreign genetic information not only from other plants but also from bacteria and bacterial viruses and other organisms. In nature, non-sexual gene transfer between plants by way of DNA transformation or viral-mediated transduction has not been demonstrated. Neither has gene transfer between bacteria and plants been demonstrated in nature although legume-*Rhizobium* N_2 fixing symbioses and bacterial-induced crown gall tumours may disprove this contention when more is learned about these systems. However, during the last 5 years several experimental demonstrations of gene transfer in cultured plant cells have been at least partially successful, thus holding the promise that introduced as well as induced variation may become a tool in plant breeding.

Doy, Rolfe and Gresshoff (1973a) used the term *transgenosis* to describe the asexual introduction of foreign genes into plant cells. They emphasize that such genes could be introduced using isolated nucleic acid or an intact virus, organelle, cell or protoplast. While they include whole plants and both animal and plant cells as potential recipients the treatment here is restricted to the transgenosis of cultured plant cells and protoplasts. The use of protoplasts as simultaneous donors and recipients (protoplast fusion) has been discussed in the previous chapter.

Isolated DNA-Mediated Transgenosis

Theoretically, isolated plant DNA is a logical choice as donor in transgenosis of plant cells. It has a better probability than bacterial or viral DNA of exhibiting homology with specific stretches of the host chromosome thus maximizing pairing and integration. The transcriptional and translational machinery of the host is probably better suited for plant DNA than for bacterial or phage DNA. However, the following transgenosis experiments point out the many technical difficulties in using isolated DNA in general and plant DNA in particular. Both plant and bacterial DNA are highly susceptible to plant cell nucleases. A plant DNA preparation in addition is much more dilute for a given gene than a bacterial preparation.

Further, transgenosis by isolated plant DNA is currently hampered by a lack of suitable, biochemically characterized selective markers. Early experiments demonstrated a poor efficiency of DNA uptake by cultured cells and protoplasts. Bendich and Filner (1971) found that only 0.5% of isotopically labelled tobacco DNA or heterologous *Pseudomonas aeruginosa* bacterial DNA was taken up into the nuclear fraction of cultured tobacco cells. The rest was rapidly degraded in spite of extensive washing and pronase treatment to reduce extra-cellular DNase activity. No evidence was obtained for the integration of this DNA into the recipient genome.

Ohyama *et al.*, (1973) suggested that protoplasts, which lack cell walls, may be better recipients of isolated DNA. In their experiments approximately 1% of exogenous *E. coli* DNA was found in an acid precipitable fraction of protoplasts of soybean, carrot and *Ammi visnaga.* By treating tobacco cells with cationic DEAE dextran, Heyn and Schilperoort (1973) increased 'uptake' of acid precipitable *Agrobacterium tumefasciens* DNA to 10–15% of input. Possibly DEAE Dextran and anionic DNA form a DNase resistant complex. However, only 0.145% of this DNA penetrated the cell; the rest was adsorbed to the cell wall.

Studies involving the selection of protoplasts and cultured cells which express a donor DNA marker have so far been few and only marginally successful. Holl *et al.*, (1974) recently described two experimental systems. One involves the acquisition of mannitol utilization by soybean protoplasts from the bacterium *Azotobacter vinelandii.* No clear-cut DNA-mediated transgenote has yet been reported. A serious difficulty in this experimental system is the impermeability of soybean protoplasts to mannitol; thus, transgenotes would have had to acquire at least two bacterial genes—a mannitol permease as well as the gene responsible for the mannitol-fructose conversion. Such double transgenotes would probably be exceedingly rare; moreover it is difficult to envisage a bacterial permease functioning in a plant cell membrane.

In the second system described by Holl and collaborators, seeds of a mutant pea strain unable to form root nodules with *Rhizobium* bacteria were treated with DNA extracted from a strain which forms functional nodules. They observed that 1–2% of those plants grown from treated seeds were corrected for nitrogen-fixation. An analysis of this result should consider the mutagenic effects of DNA and the observation that all corrected seeds simultaneously recovered two lost functions which are encoded by independently segregating genes *viz.* nodulation and N_2 -fixation. A serious restriction of this system is the limited number of seeds which can be treated. Alternatively, the authors have proposed treating cultured mutant shoot tips with DNA from the symbiosis-competent line. Transgenotes can be selected by whole plant regeneration from shoot tips on an N-free medium inoculated with *Rhizobium.* However, the cultured shoot tip consists of a heterogeneous population of partially differentiated diploid cells and it is not known if only certain types or minimal numbers of these cells must be transgenosized before the regenerated plant will fix nitrogen. As yet no transgenotes have been reported.

Ferrari, working in Widholm's laboratory (1975), has used carrot cell lines resistant (5*mt* ^{r}) and sensitive (5*mt* ^{s}) to 5-methyl tryptophan to study DNA transgenosis. The transfer of genetic information was monitored by observing the frequency of 5*mt* ^{r} clones arising among a population of 5*mt* ^{s} cells which had been treated with DNA extracted from 5*mt* ^{r} cells. Although an apparently mutagenic effect of both 5*mt* ^{r} and 5*mt* ^{s} DNA contributed to altering the phenotype of the recipient cell, intact 5*mt* ^{r} DNA was 2 to 8 times more effective than 5*mt* ^{s} DNA in conferring resistance to 5-methyl tryptophan. However, the observation that these differences occurred in only one quarter of the experiments illustrates the difficulties of this system. An understanding of the means by which input 5*mt* ^{r} DNA. effects 5-methyl tryptophan resistance awaits a biochemical comparison between donor and acquired gene products.

Obviously refinements are needed in future DNA transgenosis systems. These would include the transfer of single gene, biochemically characterized, markers to recipient cells incapable of mutational escape from selective conditions, the means of neutralizing recipient DNAases or otherwise improving recipient cell competence, and enrichment of donor (especially plant) DNA for specific genes (e.g. by fractionation, use of organelle DNA, molecular cloning or reverse transcriptase products of purified plant messengers). Further, if isolated protoplasts are best suited for DNA uptake, callus and whole plant regeneration must be achieved from protoplasts of the many species where this is not yet possible.

Viral-Mediated Transgenosis

In viral-mediated transgenosis foreign genetic information is transferred to and expressed by plant cells following exposure to intact viral particles.

The experiments described here deal exclusively with bacteriophage infection although the productive infection of plant protoplasts with plant viruses is, in a strict sense, also viral mediated transgenosis.

In a series of experiments Doy, Gresshoff and Rolfe (1973b) took advantage of the inability of tomato callus to survive with either lactose or galactose as the sole carbon source. In one experiment, high titre preparations of the specialized transducing bacteriophages ø80 and *l* were applied to tomato callus.

The former (ø80 *plac* $^+$) carried the bacterial gene coding for *b*-galactosidase, the enzyme which catalyses hydrolysis of lactose to glucose and galactose. The latter (*l pgal* $^+$) carried the bacterial structural genes encoding the enzymes which catalyse the conversion of galactose to utilizable glucose 1–P. Eighty per cent of the calluses treated simultaneously with both phages or with ø80 *plac* $^+$ alone were able to survive and grow slowly on lactose medium as compared with 5–10% of untreated controls.

In another experiment galactose conversion was shown to be dependent on intact bacterial *gal* genes since bacteriophage with no *gal* genes (ø80) or with a mutated gene for galactose utilization (*l pgal* $^-$) did not elicit growth with galactose as sole carbon source in the inoculated calluses.

More convincing molecular proof of bacterial gene expression by the treated plant cells was the immunological identification of the *b*-galactosidase activity in the lactose utilizing callus. Antiserum specific for the *E. coli* enzyme afforded the same degree of protection against thermal inactivation of the *b* -galactosidase activity in extracts of both *E. coli* and the lactose utilizing callus. Doy, Gresshoff and Rolfe (1973a,b) have emphasized that the definition of transgenosis implies neither a specific mechanism of viral transduction of bacterial genes nor the integration and stable transmission of the acquired genes.

Table . *b* -Galactosidase activity from crude extracts of haploid tomato callus and its protection by antiserum specific for *E. coli* *b* -galactosidase (Doy *et al.*, 1973b.)

Callus type	Carbon source	Growth	*b* - galactosidase specific activity	% activity remaining after 20 min. at 60°C in presence of rabbit anti-serum specific for *E. coli* *b* - galactosidase
non treated	sucrose	rapid	158	5%
non treated	lactose	none	264	5%
non treated	lactose	slow (rare)	144	5%
treated with ø 80 *plac and l p gal*	lactose	slow	6850	50–60%
non treated plus *E. coli* extract				
containing *b* - galactosidase	sucrose	fast	*	50–60%

* Adjusted to levels comparable to transgenosized cells.

Carlson (1973b) demonstrated the appearance of two early phage enzymes upon 'infection' of barley protoplasts with the lytic bacteriophage T3. The phage exposed protoplasts synthesized S-adenosyl methionine cleaving enzyme (SAMase) and RNA polymerase, two T3-encoded enzymes not normally produced by the plant.

The latter enzyme was definitely assigned a bacteriophage origin since exposure to T_3 containing an amber (nonsense) mutation in the RNA polymerase structural gene resulted in significant protoplast synthesis of SAMase activity only.

It is interesting that the phage-borne bacterial *gal* mutation introduced into tomato callus (Doy *et al.*, 1973b) was also amber, providing circumstantial evidence that plants as well as bacteria may recognize the amber codon as a stop signal. Doy *et al.*, (1973b) demonstrated that both tomato and *Arabidopsis thaliana* calluses die quite rapidly upon exposure to ø80 phage carrying a bacterial amber suppressor. It is, however, difficult to distinguish whether the bacterial suppressor tRNA interferes with termination of translation or simply competes with plant tRNA at sense codons.

It has been demonstrated (Johnson *et al.*, 1973; Grierson *et al.*, 1975) that sycamore cells in suspension can acquire the ability to utilize lactose from *l* phage carrying the *lac* genes. Ordinary *l* does not confer lactose utilizing ability. However, biochemical evidence for the presence of bacterial enzyme in cells growing on lactose has not yet been obtained.

The *lac* $^+$ sycamore cell lines as well as the lactose-utilizing tomato callus lost their new function in succeeding generations (Doy, Gresshoff & Rolfe, 1973b; Grierson *et al.*, 1975). Polacco *et al.*, (1974) demonstrated the expression of *l* borne *lac* genes in tobacco diploid callus both by growth and biochemical criteria. However, these transgenosized lines uniformly lost all *lac* functions after several months. These results demonstrate the transient nature of bacterial and viral gene expression in plant cells.

We may tentatively summarize the advantages and drawbacks of viralmediated transgenosis. The genome size of a transducing phage is small ($l = 31 \times 10^6$ daltons) so that the desired bacterial gene is enriched about 80 fold compared with its concentration in the *E. coli* genome.

Therefore, its chances of uptake and expression are improved compared with isolated *E. coli* DNA transgenosis. Further, the phage coat may serve to protect the foreign DNA from nuclease attack and even provide some transport function.

On the negative side the bacterial genes introduced probably lack homology with the plant genome thereby minimizing the probability of chromosomal integration and stable transmission. Of course it cannot be assumed that the transcription and translation of the bacterial genes by the plant host are efficient. Presumably, many phage specific genes are expressed in the recipient plant cell and this may be related to the generally deleterious effect of phage treatment on callus growth. The systems employed were far from ideal in that T3 genes conferred no selective advantage on barley protoplasts while calluses of tomato (Doy *et al.*, 1973b; Table) and tobacco (Polacco *et al.*, 1974; Fig. 17.3) and sycamore cell suspensions (Grierson *et al.*, 1975) have been shown to have plant-specific *b* -galactosidase activity.

As a result the selective pressures in the *l plac* system were probably not great; Doy *et al.*, (1973b) reported that 5–10% of control calluses grew on lactose. The presence of a homologous plant enzyme also complicates the identification of the introduced gene.

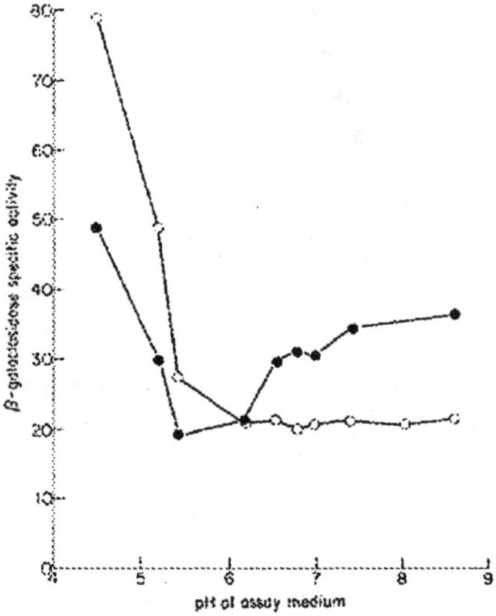

Figure: The presence of plant specific* b *-galactosidase activity in extracts of tobacco callus grown on sucrose (- -) and in extracts of callus exposed to lactose (- -) for two weeks as sole carbon source. (From Polacco* et al., *1974.)

Further experiments should utilize biochemically characterized markers which confer a more rigorous selective advantage. Whole plant regeneration should be attempted under selective conditions and modes of gametic transmission of the desired trait studied.

Ideally, to improve homology (and therefore integration), plant rather than bacterial genes should be introduced. In viralmediated transgenosis the problem is to incorporate specific plant DNA stretches in a phage coat.

A potential system would be the incorporation of chloroplast DNA in *l* phage by techniques recently described (Thomas *et al.*, 1974); such phage could possibly be used to transgenosize albino callus to green.

By involved biochemical and genetic manipulation Thomas *et al.*, prepared *l* DNA (*l gt*) with a 27% deletion. It has lost the ability to lysogenize *E. coli* but can still replicate, produce coat protein and lyse its host.

However, because the *l gt* DNA is smaller than normal it cannot associate with the coat protein in cell lysates to form an infective particle. The *l gt* DNA, therefore, cannot form plaques in transfection experiments, unless DNA from apparently *any* source, is biochemically inserted into the genome. Such DNA is replicated along with *l* DNA. Cloning such DNA-rescuėd phage is equivalent, then, to cloning a small stretch of the foreign DNA incorporated in the *l* genome.

CONCLUSION

This chapter has reviewed recent advances in the development of techniques for modifying the genetic information of cultured cells of higher plants. We have not considered nuclear and organelle uptake by protoplasts or protoplast fusion as means of introducing genetic variation (Chaleff & Carlson, 1974).

The selective systems would be similar to at least some of those already described and the techniques involved. It is hoped that all these experimental approaches will prove useful in the elucidation of the biological processes of higher plants. These procedures also may contribute to the development of improved crop varieties.

However, successful realization of these aims will require a fuller development of the systems and methods which have been described. The differentiation and chromosome composition of cultured cells must be better controlled and the minimum cell density required for growth must be reduced.

In addition, these *in vitro* techniques must be extended to a greater number of crop species.

We should not close this chapter without mention of the potential biohazards of gene transfer experiments. The experimenter should consider potential dangers to agriculture, the ecosystem and human beings.

We shall draw one possible scenario: a researcher succeeds in recovering intact infective bacteriophage from viral transgenosized callus. Subsequently the virus is found able to infect whole plants and cause severe disease symptoms and necrosis—the latter due possibly to a bacterial toxin gene or a plant cellulase gene, etc. The authors believe, however, that with proper precautions such experiments have the future potential of yielding new and valuable agricultural varieties.

7

RESISTANCE OF PLANT DISEASES

Plants are constantly confronted with a wide variety of potential pathogens within their environment. Nevertheless, the development of disease is the exception rather than the rule, due to the highly efficient nature of the coordinated systems of passive and active defences that have evolved in plants Link to Passive Defenses and Active Defenses.

These defences generally limit the host range of a microorganism capable of causing plant disease to the members of a single plant genus or species, and affect all races of a particular pathogen similarly. However, plant disease resistance can also be induced in specific plant cultivars within the host range, or only in response to specific races of pathogen.

The specificity of plant responses to pathogens can be classified into two broad categories. Non-specific resistance (general, non-host or basic resistance) is a response to all races of a particular pathogen, and occurs in all cultivars of a host plant species.

In contrast, specific resistance is dependent upon the presence of a particular pathogen race, a particular host plant cultivar, or both. The underlying genetic basis of each type of plant disease resistance differs according to the genetic makeup of both plant and pathogen.

NON-SPECIFIC PLANT DISEASE RESISTANCE

Non-specific plant disease resistance is multi-component, relying upon a foundation of passive plant defences Link, and usually also involving the activation of active defences by non-specific elicitors of biotic origin Link. The combination of defences involved in this type of resistance is highly coordinated and similar for all plant-pathogen interactions.

However, substantial variation in both the timing and degree of the active component of plant defence and in environmental factors have been shown to be critical to its success. Whilst stronger, more timely non-specific defence responses are responsible for many incompatible plant-pathogen interactions, weaker non-specific defence responses are often overcome by the pathogen in compatible interactions, and have also been observed in symbiotic relationships with endophytic fungi Link.

The genetic basis underlying non-specific plant disease resistance is complex, and involves multiple genes that encode proteins with a diversity of functions in both partners of the plant-pathogen interaction.

These can be divided into pathogenicity genes that determine the ability of the pathogen to cause disease and defence-related genes that enable the plant host to execute defence responses. Both classes of genes contain members that are expressed in a constitutive manner, in addition to genes that are only expressed in response to the interaction of plant and pathogen. Pathogen genes that govern pathogenicity usually encode proteins that have a negative impact on disease resistance.

The majority of pathogenicity genes in plant pathogens condition the ability to establish infection, and these include genes that encode proteins with specific roles in adhesion to the plant surface, the formation of penetration structures Link, cell wall degradation, and the synthesis of toxic compounds.

However, a number of pathogenicity genes instead govern the ability of the pathogen to defeat plant defences, such as those encoding proteins involved in the detoxification of phytoalexins Link. Plant defence-related genes encode proteins that enable the detection of non-specific elicitors and the activation of an intracellular signalling pathway leading to plant defences, as well as those themselves involved in passive and active plant defences.

SPECIFIC PLANT DISEASE RESISTANCE

In contrast to non-specific resistance, specific plant disease resistance appears to be governed by a single gene or a small number of related genes, which encode proteins capable of altering the outcome of an otherwise compatible plant-pathogen interaction. Genes conditioning host-pathogen specificity are found in particular subpopulations of the pathogen, plant host, or both interacting organisms, and specific plant disease resistance can be subdivided into three major categories on this basis.

Race-specific Resistance

Race-specific resistance is induced in response to only a particular race of pathogen, but occurs in all cultivars of the host plant. This type of specific disease resistance is dependent upon genetic variation within the pathogen species, and the production of proteins capable of altering the outcome of an otherwise compatible plant-pathogen interaction in only certain pathogen races.

Cultivar-specific Resistance

Cultivar-specific resistance is activated only in a specific host plant cultivar, but in reaction to all races of a pathogen species. In a few plant-pathogen systems where non-specific resistance limits the host range of the pathogen to a plant genus, this type of resistance occurs at the level of the host plant species, and is termed species-specific resistance. Cultivar-specific or species-specific resistance relies upon genetic variation within the host plant species or genus, and the production of proteins capable of altering the outcome of an otherwise compatible plant-pathogen interaction in only certain plant cultivars or species.

Race-cultivar-specific (gene-for-gene) Resistance

If both pathogen and host specificity are involved, plant disease resistance is termed race-cultivar-specific resistance, since it results only from the interaction of a particular pathogen race with a particular cultivar of the host plant. This type of resistance is usually referred to as gene-for-gene resistance, because in most cases it requires the presence of both a race-specific avirulence (avr) gene in the pathogen and one or more corresponding cultivar-specific resistance (R) genes in the host plant.

Resistance is only induced when a plant cultivar in possession of a specific resistance (*R*) gene recognises a pathogen race that contains the corresponding avirulence (*avr*) gene (a). The absense of either the avirulence gene (b), the resistance gene (c) or both (d) from the interacting organisms leads to lack of recognition by the host plant and the onset of disease.

Avirulence and resistance genes are usually dominant genes, which may exist within multigene families, and undergo a high rate of mutation in response to the presence of each other.

Maintenance of detrimental avirulence genes is thought to be due to a small, additive, pleiotropic pathogenicity role in the pathogen. In a small number of cases of race-cultivar-specific resistance that involve the production of host-selective toxins, race-cultivar-specific resistance relies upon the absence of either a race-specific gene conditioning toxin production and a cultivar-specific gene governing toxin sensitivity.

INTERACTIONS INVOLVING TOXIN PRODUCTION AND SENSITIVITY GENES

Disease only occurs in interactions involving a pathogen race in possession of a toxin-production (*TP*) gene and a plant cultivar that contains the corresponding toxin-sensitivity (*TS*) gene (a). The abssense of either the toxin-production gene (b), the toxin-sensitivity gene (c) or both (d), result in plant disease resistance.

Both types of race-cultivar-specific resistance are dependent upon genetic variation within both the pathogen and host species, and the production of proteins by only particular pathogen races and host cultivars, that are capable of acting in combination to alter the outcome of a plant-pathogen interaction. An otherwise compatible interaction results in resistance due to the presence of both avirulence and resistance genes, whereas genes conditioning toxin production and toxin sensitivity cause disease in an otherwise incompatible interaction.

MECHANISMS UNDERLYING SPECIFIC PLANT DISEASE RESISTANCE

The biochemical mechanisms responsible for the induction of specific resistance in plant-pathogen interactions are poorly understood, but are likely to vary with both the type of specific resistance and the plant-pathogen system involved. The three most common mechanisms underlying specific resistance appear to be race-specific elicitors, host-selective toxins and race-specific suppressors, but others, as yet unknown, may also exist.

Race-specific Elicitors

The majority of cases of race-specific resistance appear to result from the generation by a pathogen of race-specific elicitors of active plant defences Link, and the recognition of these by the plant host. Resistance with this biochemical basis is often also cultivar-specific (and thus gene-for-gene), since the elicitor interacts with a corresponding plant receptor that is usually unique to a particular cultivar of the host plant.

The recognition of the elicitor by its receptor is proposed to occur at the plant plasma membrane for most fungal pathogens, and within the plant cell for bacterial and viral pathogens. In bacterial biotrophs such as *Xanthomonas* and *Pseudomonas*, which are extracellular plant pathogens, this event is dependent upon a bacterial membrane transport protein that delivers the elicitor into the plant cell, and is encoded by the *hrp* gene complex Link.

Interaction of the elicitor and receptor activates a complex signal transduction pathway resulting in the induction of plant defences against pathogen races harbouring the elicitor. The elicitor is generally the protein that is encoded by the avirulence gene, however in some plant-pathogen interactions the elicitor has been found to be the product of a reaction catalysed by this protein.

Resistance genes in some cases directly encode cultivar-specific receptors of race-specific elicitors, and in these cases a direct physical interaction between avirulence and resistance gene products may occur. However resistance proteins are more likely to function by registering interactions between the elicitor and an unknown target protein, or act as unique links in the signalling pathway leading to active plant defences.

Host-selective Toxins

Race-cultivar-specific pathogen resistance can also occur due to the production of compounds that are toxic to plants. These host-selective toxins (HSTs) are generated in a race-specific manner, mainly by necrotrophic species of the fungal genera *Alternaria* and *Cochliobolus*.

A few of the approximately twenty known host-selective toxins are proteins or peptides that are directly encoded by race-specific pathogen genes. However, most are non-protein compounds of low molecular weight that are synthesised in reactions catalysed by proteinaceous race-specific gene products.

Following transport into the host plant cell via a highly coordinated delivery system, host-selective toxins cause cellular damage, but only in toxin-sensitive cultivars that harbour a single gene conditioning toxin sensitivity. The mode of action of host-selective toxins is highly variable, but appears to always involve either activation or inhibition of a cultivar-specific protein. For example, T-toxin, produced by *C. heterostrophus*, serves to activate a cultivar-specific protein capable of forming destructive membrane pores, whereas HC-toxin from *C. carbonum* inhibits a cultivar-specific version of an enzyme that modifies DNA-bound proteins to cause disturbances in gene expression.

Race-specific Suppressors

Race-specific resistance can also result from pathogen production of race-specific suppressors that inhibit a non-specific resistance response. To date, race-specific suppressors have been described for only a few species of biotrophic plant-pathogenic fungi, including *Phytophthora infestans*. In contrast to race-specific elicitors, they are proposed to interfere with elicitor binding, signal transduction, gene expression or plant defences to suppress the non-specific resistance response towards races that harbour them.

Race-specific suppressors may be proteins directly encoded by pathogen genes governing race-specificity, or may be non-protein compounds produced by reactions catalysed by these proteins. Plant disease resistance in cases involving race-specific suppressors may or may not also be cultivar-specific (and thus gene-for-gene), depending on whether or not the actual mechanism of suppression involves a cultivar-specific plant molecule.

THE RELATIONSHIP BETWEEN NON-SPECIFIC AND SPECIFIC RESISTANCES

The connection linking non-specific and specific resistances is rarely considered and thus poorly understood. However, it is likely that the relationship between these two forms of resistance differs depending on the particular type of specific resistance and the biochemical mechanism involved in its induction.

Specific resistance conditioned by race-specific elicitor molecules is generally of a higher intensity and more successful than non-specific plant disease resistance, and also involves an oxidative burst that is diphasic in nature.

These observations indicate that this type of specific resistance may be composed of a race-specific component to resistance, which is supplementary to and superimposed upon an unsuccessful non-specific resistance response. A common signalling pathway leading to defence responses may be activated to different degrees by both non-specific elicitors and race-specific elicitors resulting from the same plant-pathogen interaction.

The highly conserved nature of plant resistance mechanisms and the observation that different pathogen races can elicit defence responses that differ only in intensity both support this hypothesis. However, different pathogen races can also induce resistance responses that contain unique defence components. For example, the hypersensitive response Link is rarely induced by non-specific elicitors, but almost always occur during a race-specific resistance response.

This indicates that different signalling pathways, possibly convergent at some point, may be activated by non-specific and race-specific elicitors.

RESISTANCE AND RACE-SPECIFIC ELICITORS

Specific resistance conditioned by host-selective toxins or race-specific suppressor molecules by definition cannot involve the elicitation of a supplementary resistance response. Specificity of resistance in these cases must therefore involve modification or negation of an otherwise successful non-specific resistance response that occurs in all pathogen subpopulations and all host cultivars. The presence of race-specific suppressors in some pathogen races could be expected to prevent or lessen the non-specific defence response, either in its entirety by interfering with elicitor binding or an early signalling step, or in part by obstructing a late signalling step or an actual defence process.

RESISTANCE AND RACE-SPECIFIC SUPPRESSORS

In contrast, the activity of host-selective toxins is probably superimposed on a non-specific resistance response occurring in its entirety, and merely lessens its influence by killing host plant cells before their defences can be fully activated.

Plant disease resistance is a complex phenomenon that most commonly occurs in a non-specific manner, as a result of multiple genes conditioning the ability of the pathogen to cause disease and enabling the plant host to mount an effective defence response. However, plant disease resistance can also be induced only in response to particular pathogen races (race-specific resistance), only in particular host plant cultivars (cultivar-specific resistance), or only when both a specific pathogen race and plant cultivar interact (race-cultivar-specific or gene-for-gene resistance). This host-pathogen specificity can be attributed to a single gene or a small number of related genes enabling the production of race-specific elicitors, host-selective toxins, or race-specific suppressors in different host-pathogen systems.

Specific plant disease resistance resulting from race-specific elicitors is probably superimposed upon a non-specific resistance response that has been overcome in the host range of the pathogen. In contrast, host-selective toxins and race-specific suppressors most likely achieve host-pathogen specificity in disease resistance through the modification or negation of an otherwise successful non-specific resistance response.

PLANT DEFENSES AND DISEASES

Plant pathogens fall into two broad categories: necrotrophs (those that kill plant cells before parasitising them), and biotrophs (those that obtain nutrients from living cells). Failure of pathogens to invade suitable host cells (dead in the case of necrotrophs; alive for biotrophs) will prevent them from infecting the host and the plant will be *resistant*. Additionally, the establishment of a parasitic relationship is dependent on the responses of the plant under attack.

Resistant hosts prevent or slow the development and reproduction of the majority of pathogen propagules that they come into contact with. Resistance can be expressed at many stages in the infection process, from inhibition of propagule germination and penetration, to the restriction of colony development after the pathogen has become established. The defence barriers erected by plants are a co-ordinated system of molecular, cellular and tissue-based responses to pathogen attack.

PASSIVE DEFENSES

Protection from a pathogen's initial invasion is achieved via passive defences, such as physical and/or chemical barriers. Physical barriers largely involve properties of the plant surface, that is, the cuticle, stomata and cell walls. Pathogens produce a range of cutin-degrading enzymes, which are often crucial to the successful penetration of the plant tissue.

The thickness of the cuticle, the presence of secondary cell wall, and the size of stomatal pores can all affect the success with which a pathogen invades a host. Some plants invest in very thick walls and/or cuticles, and bark (where present) can also provide a physical impediment to infection. The vertical orientation of leaves can also add to plant resistance, by preventing the formation of moisture films of the leaf surfaces, inhibiting infection by pathogens reliant on water for motility.

Chemical barriers include compounds, such as "phytoanticipins", that have antimicrobial activity and compounds that affect the vectors of plant viruses. Phenols and quinones are two classes of antimicrobial compounds produced by some plants Inhibiting compounds may be excreted into the external environment, accumulate in dead cells or be sequestered into vacuoles in an inactive form. The young fruit of numerous plants (e.g. mangoes, avocado) contain antifungal or antimicrobial compounds that are gradually metabolised during fruit ripening, making unripe fruit less susceptible to disease than ripe fruit. Lactones, cyanogenic glucosides, saponins, terpenoids, stilbenes and tannins are also plant-produced compounds associated with pathogen resistance. Saponins are a class of phytoanticipins that destroy membrane integrity in saponin-sensitive parasites, and which are stored in an inactive form in the vacuoles of the plant cell, becoming active when hydrolase enzymes are released following wounding or infection.

Some pathogens are able to release enzymes that detoxify plant saponins, making them insensitive to this line of defence. Conversely, resistance of some plants to specific pathogens is the result of an insensitivity to pathogen-produced host specific toxins. Resistance genes may encode an enzyme that converts the toxin into a non-toxic derivative or the absence of a receptor to the toxin. Another group of defensive compounds are the plant defensins, which interfere with pathogen nutrition and retard their development. Secreted defensins can create an antimicrobial microenvironment for germinating seeds and accumulated defensins can provide defence against insect-transmitted viruses in flowers, leaves and tubers. There are also proteins, both constitutive and induced that play a role in plant defence.

ACTIVE DEFENSES

Pathogen Recognition

Plants use a vast array of signals originating from micro-organisms and the environment to recognise pathogens and elicit plant defence responses. Non-specific elicitors of biotic and abiotic origin induce host defences in a broad range of host species.

Abiotic elicitors such as heavy metal ions or UV light can induce stress responses in exposed tissues, which may provide an additional barrier to invading pathogens or alternatively, increase the plant's susceptibility to infection.

Biotic elicitors include cell wall fragments released from fungi and bacteria, hydrolytic enzymes of plant or pathogen origin, some peptides, glycoproteins and polyunsaturated fatty acids. These elicitors induce defence responses in a range of host species. Often, non-specific elicitors act as a general indication that the cell has been damaged in some way (for example, the release of fragments of the host's own cell wall can elicit defence responses).

Specific elicitors enable defence against a very specific pathogen, and are conditioned by *avirulence genes in that pathogen*. Avirulence genes determine the pathogen's host range, but are only able to function in the presence of another set of genes, the 'hypersensitive response and pathogenicity' (Hrp) gene cluster. Some Hrp gene products are involved in disguising the pathogen from host recognition, thus playing a role in both virulence and avirulence.

For a biotroph to form a successful infection, it must establish a basic compatibility with its host. The pathogen may also produce compatibility factors that delay, avoid or negate recognition by a normally resistant host plant.

Virulent strains appear to be able to suppress the resistance mechanisms of the host, but are not able to halt resistance responses once they are activated. Incompatibility between a host and a pathogen results in the recognition of the pathogen and activation of defence mechanisms, while compatibility results in infection.

Specific elicitors are encoded by avirulence genes, and these peptides are believed to bind to receptor peptides, encoded by host resistance genes. Recognition of the avirulence gene products by the host triggers signal transduction pathways that cause a massive shift in gene transcription and plant cell metabolism, and local and systemic signals are released that prime the rest of the plant against further infection. The presence of non-specific elicitors, such as the release of host and pathogen wall fragments, during this process may amplify the defence response.

Host-parasite specific resistance is determined by the interaction of between products of pathogen avirulence genes, Specific elicitors and products of host resistance genes. The defence responses of plants can be very rapid. Host gene expression begins within minutes, or even seconds, of exposure to elicitors or pathogens. A diverse range of elicitors can induce a common set of responses in the host, suggesting that second messengers are involved in the signalling pathway between pathogen attack and host response.

RAPID ACTIVE DEFENSES

Almost every host-parasite interaction is unique in the details of the activation, localisation, timing and magnitude of the defence responses.

At the Membrane

The host membrane appears to be involved in the earliest stages of pathogen recognition and signal transduction. A change in membrane permeability after exposure to a pathogen causes fluxes in ions, such as K+, H+ and Ca2+, and results in changes to gene activation and the triggering of the defence responses.

Also at the membrane, the 'oxidative burst', which involves the generation of reactive oxygen species, such as hydrogen peroxide, triggers signals that affect gene expression, cross-linking in the host cell wall and initiation of later defence responses. The reactive oxygen species at the site of infection are also produced in quantities capable of killing micro-organisms directly.

At the Cell Wall

Preparations for the reinforcement of the cell wall, which can improve host resistance, begin very quickly after a pathogen attempts to penetrate a host cell. This is characterised by an intensification of cytoplasmic streaming and the accumulation of host cytoplasm around the site of attempted penetration. The cytoplasmic aggregates are thought to contain cellular apparatus for the synthesis of cell wall fortifications.

If the host cell can repair and reinforce its cell walls quickly enough, it might reduce the penetration efficiency of the pathogen. Several types of reinforcement are produced by host cells. A papilla is a deposit of callose, silicon, lignin and proteins between the cell wall and cell membrane, directly below the point of attempted penetration, while lignitubers are lignified callose reinforcements that ensheath invading hyphal tips.

Lignitubers

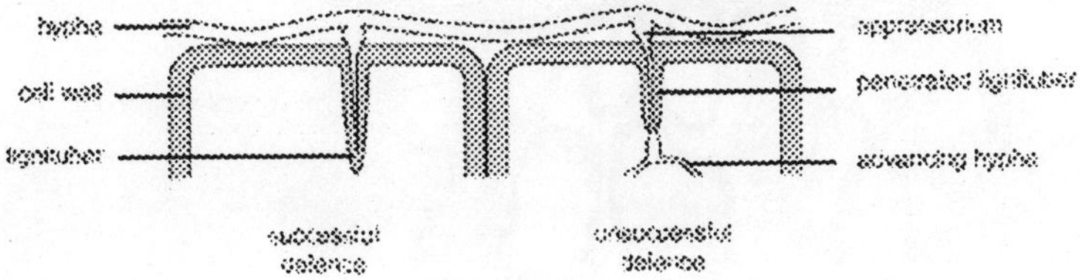

This figure illustrates the difference between successful and unsuccessful defense against a fungal pathogen. In successful defense, the lignituber contains the fungal hypha, preventing further growth. In unsuccessful defense, the hypha is able to penetrate the lignituber and continue its growth into the plant.

Hydroxyproline-rich glycoproteins are structural cell wall proteins involved in secondary cell wall thickening. The expression of genes governing their production is activated ahead of invading hyphae, reinforcing walls.

Cross-linking of hydroxyproline-rich glycoproteins caused by the release of hydrogen peroxide in the oxidative burst also reinforces cell wall compartments. Rapid deposition of lignin and suberin following infection also increases resistance to pathogens in many plants.

Lignin can also bind to hyphal tips and bacteria, physically restraining them and restricting the diffusion of their enzymes and toxins into, and the extraction of water and nutrients out of, the host cell. Cell wall reinforcements tend to be larger and more quickly formed in resistant hosts than in susceptible hosts, and inhibition of the production of callose or lignin synthesis by the pathogen enhances its penetration efficiency.

THE HYPERSENSITIVE RESPONSE

Hypersensitive Cells

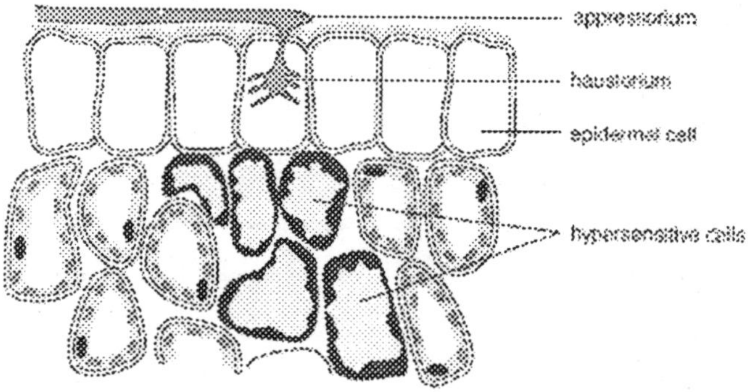

This diagram shows the penetration of an epidermal cell by the appressorium of a fungal pathogen. The hypersensitive cells respond to the invasion by dying in order to prevent further spread of the pathogen.

Hypersensitive cell death is another widespread mechanism used by hosts to prevent the spread of a pathogen. Infected cells and those surrounding them "suicide", preventing further spread, and in some cases, killing the pathogen.

It is often associated with the initiation of other responses, such as lignification and the synthesis of anti-microbial compounds. The success of hypersensitive cell death as a resistance mechanism depends on the nutritional requirements of the specific pathogen and the timing, magnitude and location of the host response.

Antibiotic Compounds

Phytoalexins are low molecular weight antibiotics produced by many (but not all) plants in response to infection. There are many biotic elicitors of phytoalexin production, such as cell wall components, as well as abiotic elicitors, such as heavy metals and ultraviolet light. Phytoalexins inhibit the growth of bacteria and fungi *in vivo* and *in vitro*, and production of these antibiotics during an infection can induce resistance to subsequent infections by that pathogen. Over 350 phytoalexins are known in over 100 plant species.

They include pterocarpans, sesquiterpenes, cryptophenols, isocoumarins, isoflavenoids, and others. Phytoalexins may be produced by any part of the plant, although different phytoalexins can accumulate in different organs.

Generally, related plant species produce structurally-related phytoalexins, and many produce more than one, enabling the plant to present a toxic cocktail to invading pathogens. Phytoalexins are produced in cells surrounding an infection site and delivered to the infected cell packaged in lipid vesicles, creating a toxic micro-environment in the infected cell and, hopefully, preventing disease establishment.

Phytoalexin accumulation is often associated with hypersensitive cell death, although only living cells can synthesise phytoalexins. Some plants can also sequester phytoalexins into vacuoles as stores of inactive sugar-conjugates, which can be cleaved and released quickly if initial defence responses are unsuccessful.

DELAYED ACTIVE DEFENSES

Delayed active defences include containment of the pathogen, wound repair, expression of pathogenesis-related proteins and the acquisition of systemic resistance. These mechanisms restrict the spread of the pathogen after infection is established and contain the damage to host tissues.

Physical Responses

The ability to repair wounds can help protect the plant from further infection by other, opportunistic pathogens. A secondary meristem in fleshy tissues, fruits, roots and bark, the cork cambium, can produce cork cells, which have thick, suberised walls. These cells can create a barrier to further colonisation by the pathogen and, in some cases, develop an abscission layer around the site of infection, causing the infected tissue separate from the healthy tissue. Wounded tree trunks often secrete protective gums that seal the wound from further infection.

The formation of tyloses, ingrowths if the protoplasm in xylem parenchyma can also restrict the spread of pathogenic propagules in the xylem, although they also tend to reduce the movement of water through the vessels, causing water stress in the plant.

PATHOGENESIS-RELATED PROTEINS

Protein Synthesis

There is a range of novel proteins synthesised in response to infection, many of which have b-glucanase, chitinase or lysozyme activity. Some pathogenesis-related proteins disrupt pathogen nutrition.

The presence of low levels of these proteins in healthy plants suggests that they might have other roles in plant growth and development aside from disease resistance. Chitinase and glucanase accumulate in the vacuoles, and glucanase is also sometimes secreted into the intercellular space.

They dissolve the fungal cell wall, fragments of which then elicit hypersensitive cell death. The breakdown of the vacuole during decompartmentalisation of the cytoplasm results in a flood of hydrolytic enzymes, which have antiviral, antifungal and antibacterial activity. The accumulation of pathogenesis-related proteins peaks around 7-10 days after initial infection. The presence of these proteins *before* infection increases the plant's resistance to pathogens, as in the case of systemic acquired resistance.

Systemic Acquired Resistance

Systemic acquired resistance, or induced resistance, is characterised by the increased resistance of a plant to a wide range of pathogens following infection by one pathogen. It is therefore fundamentally different from the specific antigen-antibody mechanism of resistance seen in the immune response of mammals.

Rather than providing immunity *per se*, systemic acquired resistance reduces the severity of later diseases.

The development of systemic acquired resistance usually requires the development of a slowly expanding necrotic lesion and other localised responses to infection, the release of a phloem-translocated signal originating from the infection site, and the subsequent priming of the plant against further attacks, allowing a more rapid response in the case of future infections.

The nature of the signal that triggers systemic acquired resistance is as yet unknown, and is likely to be a complex signal transduction pathway mediated by a number of stress signals. Salicylic acid plays a key role, via interaction with salicylic acid-binding proteins that can cause build-up of reactive oxygen species or activate gene expression.

The levels of salicylic acid increase around necrotic lesions and remain high in plants that have acquired resistance. It is not, however, salicylic acid itself that acts as the signal that is translocated systemically throughout the plant.

BIBLIOGRAPHY

Acquaah, George: *Understanding Biotechnology: An Integrated and Cyber-Based Approach*, NJ, Pearson/Prentice Hall, 2004.

Adelberg, Edward A.: *Papers on Bacterial Genetics*, Boston, Brown and Company, 1960.

Angel, Bradley: *Toxic Threat to Indian Lands: A Greenpeace Report*, San Francisco, Greenpeace, 1991.

Bailey, Conner and Charles E. Faupel: *Environmentalism and Civil Rights in Sumter County*, Colorado, Westview Press, 1992.

Bailey, Ronald: *Liberation Biology: The Scientific and Moral Case for the Biotech Revolution*, Amherst, Prometheus Books, 2005.

Bernauer, Thomas: *Genes, Trade, and Regulation: The Seeds of Conflict in Food Biotechnology*, Princeton, Princeton University. 2003.

Brooks, Rodney A.: *Flesh and Machines: How Robots Will Change Us*, New York, Pantheon Books, 2002.

Campbell, R. R., and Wade, J. L.: *Society and Environment: The Coming Collision*, Boston, Allyn and Bacon, 1972.

Chrispeels, Maarten : *Plants, Genes and Crop Biotechnology*, Sudbury MA, Jones and Barlett Publishers, 2003.

Churchill, Ward: *Struggle for the Land: Indigenous Resistance to Genocide, Ecocide, and Expropriation in Contemporary North America*. Monroe, Common Courage Press, 1993.

Daubenmire, R. F.: *Plants and Environment*, New York, Wiley, 1947.

David Sadava: *Plants, Genes and Crop Biotechnology*, Sudbury MA, Jones and Barlett Publishers, 2003.

Dodds, John H.: *Plant Genetic Engineering*, New York, Cambridge University Press, 1985.

Erickson, Kai: *A New Species of Trouble: Explorations in Disaster, Trauma, and Community*, New York, W.W. Norton, 1994.

Freidberg, Susanne: *French Beans and Food Scares: Culture and Commerce in an Anxious Age*, New York, Oxford University Press. 2004.

Hill, David: *The Quality of Life in America; Pollution, Poverty, Power, and Fear*, New York, Holt, Rinehart and Winston, 1973.

Metting, Jr., F.B.: *Soil Microbial Ecology: Applications in Agricultural and Environmental Management*, Marcel Dekker, New York, 1992.

Murray, David: *Seeds of Concern: The Genetic Manipulation of Plants*, Sydney, University of New South Wales, 2003.

Postman, Neil: *Technopoly: The Surrender of Culture to Technology*, New York, Vintage Books, 1992.

Primrose, S.B.: *Principles of Gene Manipulation*, London, Blackwell Scientific Publications, 1989.

Shetty, Kalidas: *Food Biotechnology*, New York, Dekker/CRC Press, 2005.

Walden, Richard: *Genetic Transformation in Plants*, England, Open University Press, 1988.

INDEX

O

P

R

S

T

◆◆◆